Studies in Computational Intelligence

Volume 716

Series editor

Janusz Kacprzyk, Polish Academy of Sciences, Warsaw, Poland
e-mail: kacprzyk@ibspan.waw.pl

Robert Koprowski

Processing Medical Thermal Images

Using Matlab®

Springer

Robert Koprowski
Department of Biomedical Computer
    Systems, Institute of Computer Science
University of Silesia
Sosnowiec
Poland

Additional material to this book can be downloaded from http://extras.springer.com

ISSN 1860-949X   ISSN 1860-9503   (electronic)
Studies in Computational Intelligence
ISBN 978-3-319-87056-4   ISBN 978-3-319-61340-6   (eBook)
DOI 10.1007/978-3-319-61340-6

Foreword

Since ancient times, engineers and physicians have always found ways to repurpose new technologies for the improvement of human health.

Infrared thermography, which detects black body radiation emitted by a warm object and produces images of thermal patterns, is a case in point. Efficient sensors to detect infrared energy had their origin in the first third of the twentieth century. These led to the development of night vision goggles and other devices for military use in World War II and much more sophisticated devices for various military uses up to the present time.

After the end of WW II, thermographic imaging technology was made available for civilian use. Beginning in the late 1950s, enthusiastic papers began to appear in the medical literature concerning medical applications of thermal imaging. "Thermography ... constitutes a powerful tool for non-destructive testing which has been applied successfully, not only in situations in industry, but also in the field of medicine" one review proclaimed in 1963 in the Journal of the American Medical Association 1963 [1]. The authors of that review considered thermography useful for "physiologic studies of blood flow and oxygen consumption... the effects of various psychic states, and the influence of systematic or locally acting drugs..." including the detection breast cancer.

Detecting (either screening or diagnosis of) breast cancer became the poster child for thermography. By the 1960s, physicians came to realize the medical importance of screening for breast cancer, and (X-ray) mammography began to be used for large screening programs in the late 1960s. While mammography could detect a tumor before it was palpable by physical examination, it had obvious disadvantages, not least of which was discomfort to the patient from the required breast compression. Some experts at the time estimated that X-ray mammography had the potential of creating more breast tumors (due to exposure to ionizing radiation) than it detected.

Envisioning a huge prospective market for thermographic breast imaging, major companies such as General Electric began to develop infrared cameras for breast imaging. John Webster, in the 1978 edition of his famous biomedical engineering textbook *Medical Instrumentation* [2], includes the photograph of a woman having her breasts imaged by an infrared camera. Judging by the number of research papers in the medical literature [3], the interest of the medical community in thermographic breast imaging crested in the mid-1970s. But doctors generally lost interest in the method after several clinical studies appeared later in that decade that showed that thermography was insufficiently reliable for detecting breast cancer. Webster dropped the photograph in later editions of his book. However, the technology improved with time, in terms of image quality and thermal sensitivity, and some research groups continued to pursue breast imaging. Nevertheless, a systematic review in 2013 [3] found "insufficient evidence" for the effectiveness of digital infrared thermography for breast cancer screening and diagnosis—after four decades of research on the topic.

Within the past decade, there has been a resurgence of studies on thermography for breast cancer detection. Many of these are engineering studies by groups from around the world in the hopes of improving the diagnostic performance of thermography for breast cancer screening. But many other applications of thermography are being pursued as well; some of which are described in this book.

In addition to its use in mainstream medicine, chiropractors and alternative medicine practitioners are aggressively promoting infrared thermography. Many alternative medicine clinics offer breast imaging, with claims that it can detect future propensity to develop cancer. The American College of Clinical Thermology (a group with a strong alternative medicine focus that promotes the technology) lists 74 "Indications for Thermographic Evaluation" on its website (http://www.thermologyonline.org/Patients/patients_indications.htm). These range from "altered ambulatory kinetics" to "whiplash." The level of clinical evidence for these indications varies greatly.

This volume concerns signal processing methods for thermographic images. It is surely important to get the most reliable information one can from thermographic images, and image processing is a valuable tool for that end. Careful measurements and analysis of data is central to any reliable assessment of the technology for clinical uses and for development of new applications, and this book offers practical tools for such purposes.

But engineers and doctors inhabit different universes, in one case the world of mathematics and signal processing and in the other the world of randomized clinical trials, systematic reviews, and evidence-based practice guidelines. Engineers need to work closely with physicians to identify medically important problems and medically feasible solutions. They cannot do it by themselves. In the end, one must prove that an innovation actually improves patients' outcomes. Some do and some do not, and the key to success is to choose new applications that have a chance of success and to avoid spinning ones' wheels on projects that are unlikely to succeed in medicine.

In the aftermath of the 1970s debacle about thermographic breast imaging, Moskowitz and colleagues wrote a wise paper for engineers [4], outlining the first steps involving tiny clinical trials that should be taken to assess whether a new thermographic method is worth pursuing (as opposed to showing that a method actually works, which requires a large clinical trial). Their advice is still important.

And do not forget to get clearance from the local ethics board for any studies involving human experimentation.

There is no shortage of potential research topics in medical thermography. The costs of entry into this field are vastly lower than when the field got started; a decent infrared camera can be purchased as an add-on to a smartphone on eBay for about US $100, whereas the cameras used for breast imaging in the 1970s cost US $50,000 or more in 1970 dollars. Let the cameras roll!

Prof. Kenneth R. Foster
Department of Bioengineering
University of Pennsylvania
Philadelphia, PA, USA

Preface

Modern methods of image analysis and processing are becoming more pervasive, both in private and professional life. They cover virtually every aspect of life, from industrial automation systems through building and energy industry to medicine. Medicine is also currently experiencing a boom thanks to the spread of modern equipment and electronics as well as systems of image analysis providing new quantitative diagnostic quality. One such area of medicine where an interdisciplinary approach (of an engineer and a physician) is necessary is thermal imaging, which, although known for many decades, has not been computer-assisted in any way. The use of advanced processing in thermal imaging provides quantitative data on temperature distribution and its changes in an automatically or manually separated area. In addition, modern processing methods enable to track objects and thereby stabilize their position in the scene. This monograph is therefore intended for students of biomedical engineering (provided that its price is not an obstacle), graduate students, and researchers dealing with interdisciplinary issues related to engineering and medicine. This monograph may also be of interest to IT specialists working in MATLAB[®][1] in the field of image analysis and processing as well as doctors who want to expand their knowledge on modern diagnostic methods. Finally, operators and retailers of infrared cameras interested in the possibilities and limitations of acquired image processing methods may find this monograph useful. An open-source code allows for any application of the presented methods as well as their modification and expansion. To ensure the reliability of the monograph recommendations for particular groups of readers, it should be also mentioned what aspects are particularly highlighted. Readers will not find here the history of thermal imaging, clinical results, or summary of infrared camera types. The monograph primarily attempts to show the author's individual approach to thermal image analysis and processing.

Sosnowiec, Poland   Robert Koprowski

---

[1]The word "MATLAB" will be used further in this book for simplification reasons

Acknowledgements

First, I would like to thank Dr. Sławomir Wilczyński from the Medical University of Silesia in Sosnowiec for medical consultations. I would also like to thank him for the opportunity to test the described algorithms on some selected sequences of infrared images.

I also thank Prof. Kenneth Foster from the University of Pennsylvania, USA, for consultations and interesting discussions on contemporary applications and restrictions of thermal imaging in medicine.

In particular, I would like to thank all the people who gave me the opportunity to learn the medical aspects of thermal imaging. In the area of lateral spinal curvatures, I thank Prof. Daniel Zarzycki from the Specialist Rehabilitation and Orthopaedic Centre for Children and Youth in Zakopane, Poland; Dr. Henryk Konik from the Trauma and Orthopaedic Surgery Ward of the Complex of Town Hospitals in Chorzow, Poland; and Dr. Bronislaw Kajewski from the Central Mining Institute in Katowice, Poland, and in the area of dermatology and aesthetic medicine, Dr. Slawomir Wilczyński from the Medical University of Silesia in Sosnowiec and Dr. Arkadiusz Samojedny from the Silesian Medical School in Katowice, Poland. Imaging of newborns and assessment of their comfort was possible thanks to the collaboration with Dr. Katarzyna Wojaczyńska-Stanek from the Medical University of Silesia in Katowice. I would also like to thank Paweł Rutkowski, a representative of the FLIR Company for making available the camera and providing the opportunity to test it in medical applications. Finally, I would like to thank Prof. Zygmunt Wróbel and all my colleagues from the Department of Biomedical Computer Systems of the University of Silesia, Poland, for valuable casual talks and discussions on the practical usefulness of thermal imaging in medicine.

Thank you to all those who have always been with me and supported me while overcoming various life's challenges—including writing this book.

Contents

Purpose and Scope of the Monograph

The main objective of this monograph is to present well-known as well as modified and new methods of analysis and processing of thermal images, in particular medical images. In addition, it discusses and presents parts of the source code along with the graphical user interface (GUI). The full source code is attached to this monograph and can be used by the reader without restrictions.

The monograph describes its purpose and scope:

Chapter 1 presents the state of the art on thermal imaging and image analysis as well as diagnostic problems related to thermal imaging applied in medicine.

Chapter 2 describes image acquisition. Particular attention is paid here to image acquisition and errors resulting from inadequate preparation of the patient for examination.

Chapter 3 is devoted to the methods of image preprocessing, starting with filtration, through the selection of ROI, and ending with image stabilization.

Chapter 4 presents the methods of thermal image analysis and processing, starting with analysis of temperature changes for the longitudinal or transverse profile and ending with advanced methods of image analysis and processing as well as classification.

Chapters 5 and 6 of the monograph are devoted to examples of tailoring algorithms to specific applications. In Chap 5, this is analysis based on a single image, while the seventh chapter concerns analysis of image sequences.

Chapter 7 of the monograph is practical assessment of the algorithm sensitivity to parameter changes.

Chapter 8 of the monograph is a summary of the discussed methods of thermal image analysis and processing.

The appendix includes a description of the names of m-files and functions which are attached to this monograph.

Symbols

A	Amplitude
d_{mn}	Parameter responsible for the number of analyzed pixels
G_i	Total number of iterations
G_B, G_P, G_V	Transmittance
h	Filter mask
i	Number of the image in a sequence
I	Number of images in a sequence
J_G	Criterion
k_G	Gain coefficient
$L(m,n,i)$	Point of the matrix of the image L
L_{BIN}, L_{BI}, L_{BOUT}	Binary image
L_{GRAY}	Image in gray levels
L_{DIL}	Image of dilation
L_{ERO}	Image of erosion
L_{OPE}	Image of morphological opening
L_{CLO}	Image of morphological closing
L_{ROI}	ROI
L_{gauss}, L_{gauss2}	Gauss transformation matrix
$l_{1,2,\ etc}$	Characteristic points in the image
M	Number of rows in a matrix
N	Number of columns in a matrix
m	Row
n	Column
p_r	Threshold
p_{std}	Standard deviation of the mean
SE, SEE, SED	Structural element
T	Temperature
$\ominus$	Erosion symbol
$\oplus$	Dilation symbol
β	Angle of spinal curvature

γ	Angle of analysis
τ	Time constant
$\sigma_m,\ \sigma_n,\ \sigma_i$	Standard deviation of the mean

Abstract

This book presents automatic and reproducible methods for the analysis of medical infrared images. The presented methods have been practically implemented in MATLAB, and the source code has been presented and discussed in detail. These methods have been verified with medical specialists. This book is therefore addressed to IT specialists, bioengineers, and physicians who wish to broaden their knowledge of the tailored methods of medical infrared image analysis and processing.

Chapter 1
Introduction

1.1 State of the Art

The state of the art on thermal imaging refers to the last decades. This is mainly related to dynamic development of infrared cameras in this period. A pioneer in the human body thermometry was Carl Reinhold August Wunderlich (born in 1815). Previously (in 1593), the first experiments on "reflected heat" were conducted by Jean Batista Della Porta of Naples. The next ones were conducted by William Herschel (1800), John Herschel (1840), Charles Babbage, JD Hardy (published in 1935), Samuel Langley, Marianus Czerny, Peter Timofeev and Max Cade. I encourage the readers interested in the history of infrared thermography to read the articles published by these scientists or the summary presented in [5]. Today in medicine, despite the restrictions on the form of measurements (which are presented in detail in the following chapters), thermal imaging is gaining more and more supporters. An example is a growing number of articles on thermal imaging. The results for the first 5 countries, 5 Institutions, 5 authors, 5 journals related to the word "thermal imaging" are shown in Table 1.1 (according to AuthorMapper).

Table 1.1 shows that nowadays thermal imaging is primarily applied in cardiology to evaluate and verify any type of patency of veins and arteries after any type of surgery. Other areas include assessment of the laser effect on the skin and patients in general (423 articles in Laser in Medicine Science). Most publications, 6077, are from the United States, the next country, namely Germany, has three times fewer articles (2084). There is a similar number of articles from China and Italy—about 1400. Authors writing articles on the issue of 'thermal imaging' have a maximum of 34 articles—Liang Ping, and 33—Vogl Thomas. In addition to them, attention should be paid to the other authors who have been dealing with thermal imaging almost all their lives, for example, EFJ Ring and K Ammer from the University of Glamorgan, A. Nowakowski from the Technical University of Gdansk, Poland, B. Więcek from the Technical University of Lodz, Poland and many others. However, apart from AuthorMapper, there are other databases that are the basis for judging professional

© Springer International Publishing AG 2018
R. Koprowski, *Processing Medical Thermal Images*, Studies in Computational
Intelligence 716, DOI 10.1007/978-3-319-61340-6_1

Table 1.1 The first 5 countries, 5 institutions, 5 authors, 5 journals related to the word "thermal imaging"[a]

First 5 countries	United States	Germany	United Kingdom	China	Italy
Number of publications	6077	2084	1559	1490	1427
Examples of publications	[6, 7]	[8, 9]	[10, 11]	[12, 13]	[14, 15]
First 5 institutions	University of California	University od Toronto	Harvard Medical School	Mayo Clinic	Chinese Academy of Sciences
Number of publications	154	117	109	100	97
Examples of publications	[16, 17]	[18, 19]	[20, 21]	[22, 23]	[24, 25]
First 5 authors	Liang Ping	Vogl Thomas	Gangi Afshin	Lencioni Riccardo	Ricke Jens
Number of publications	34	33	28	27	24
Examples of publications	[26, 27]	[28, 29]	[30, 31]	[32, 33]	[34, 35]
First 5 journals	Nanoscale Research Letters	Analytical and Bioanalytical Chemistry	Molecular Pain	Laser in Medicine Science	CardioVascular and Interventional Radiology
Number of publications	2168	569	494	423	371
Examples of publications	[36, 37]	[38, 39]	[40, 41]	[42, 43]	[44, 45]

[a] Status as at 10.26.2016

activity and advancement in different countries. These are, for example, PubMed or Web of Science. In accordance with the latter database, the number of citations of all articles was traced using the query: TOPIC: (thermography) AND TOPIC: (infrared). The result was 6510 articles, and the article entitled "Pulse phase infrared thermography" by Maldague and Marinetti [46] had the largest number of citations, i.e. 365. The other articles have a lower number of citations, for example, an article by Jones BF entitled "A reappraisal of the use of infrared thermal image analysis in medicine" [47]. A detailed summary of the number of publications and the total number of citations according to Web of Science is shown in Figs. 1.1 and 1.2.

It should be noted that different names relating to thermal imaging have appeared. Thermal cameras are also called infrared cameras or thermal imaging cameras. The same applies to thermography imaging, which is called Infrared thermography (IRT), thermal imaging, and thermal video. The results of infrared

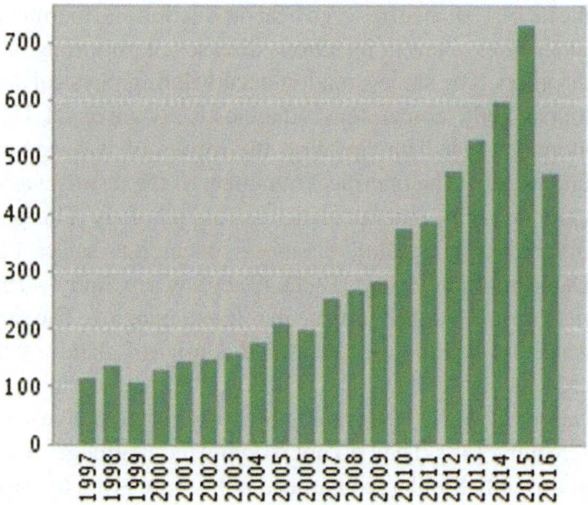

Fig. 1.1 Items published in each year (the last 20 years) (https://www.authormapper.com/)

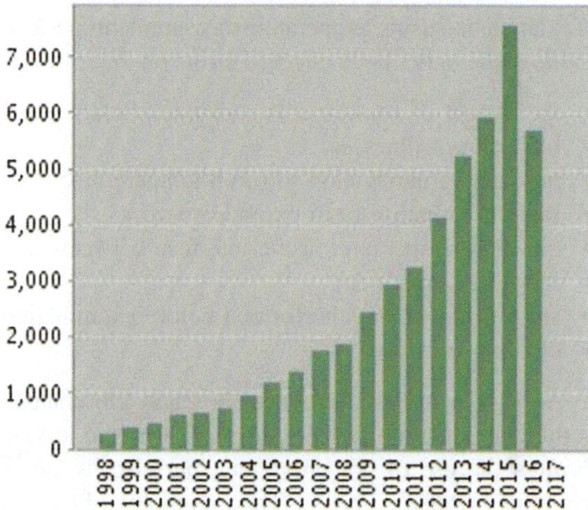

Fig. 1.2 Citations in each year (the last 20 years) (https://www.authormapper.com/)

imaging are thermal images as well as infrared images. Both terms are used alternately in this monograph.

1.2 Infrared Imaging and Diagnostic Significance

Infrared imaging certainly provides the simplest and intuitively easiest results for medical interpretation. After all, temperature measurement has been the basis for the diagnosis of various pathologies in humans for hundreds of years. The simplest interpretation is subfebrile state or fever in a patient. The body temperature of a healthy subject is from 36 to 37 °C. Values exceeding this range are associated with the lack of balance between heat production and thermoregulatory effector mechanisms. Such cases occur in infectious diseases or poisoning, and also as a result of hormonal disorders. The same situation occurs during physical activity. Probably for this reason human body temperature is diagnostically universal. Unfortunately, perhaps due to the aforementioned universality, the results of temperature measurements were treated uncritically. The uncritical treatment of the results was directly related to the interpretation of each temperature increase as pathology. This process is known, in particular, in infrared imaging and diagnosis of breast cancer (The Breast Cancer Detection Demonstration Project (BCDDP) Follow-up Study—1980). Every temperature change (its increase) looks the same in infrared imaging. The same temperature changes can be caused by a tumour or increased density of glandular tissue resulting from hormonal changes. In 2011 the U.S. Food And Drug Administration issued a safety communication saying that thermography is not an alternative to mammography in breast cancer screening [48]. They do not recommend thermography as a replacement for mammograms [48]. Similar recommendations were given by Croatian Society of Radiology and Croatian Society of Senology in 2013 [49, 50]. Not only can the diagnosis of breast cancer yield false positive results. This type of false results can occur for any other type of examination using temperature measurements performed with infrared cameras. This is due to the rather large effect of errors on the results obtained. These include:

- the impact of warm meals consumed before examination,
- undisclosed illnesses,
- lack of acclimatization to room temperature,
- thermal radiation from external sources,
- elevated temperature at the interface between skin and skin,
- fear of examination,
- sudden movements performed before examination,
- clothing pressure.

A majority of these errors can be easily eliminated. Operators and doctors can do nothing about some of them. Most known and unknown sources can be eliminated through the use of not only static but also dynamic analysis of temperature distribution (Pulse Phase Thermography—PPT, Pulsed Thermography—PT, Modulated Thermography—MT). Dynamic analysis involves sequential acquisition of a series of infrared images usually of the object that is stimulated. This stimulation can be local cooling or warming of the subject (patient). This type of dynamic analysis provides results, mostly time constants, regardless of the absolute temperature value. The resulting time constants provide much better sensitivity and specificity than in the case of analysis of the absolute temperature value. However, they are more time-consuming in terms of acquisition and analysis.

Another element, which should be emphasized here, is related to the effect of parameter changes on the results obtained. Analysis of this type is extremely important when measurements on the patient are carried out by one operator. The impact of individually set parameters in the data analysis algorithm on the results is also vital. Unfortunately, the designed data analysis algorithms must be tailored to a specific application. This leads to a contradiction. On the one hand, algorithms must be tailored (as mentioned earlier), while on the other hand they should be universal enough to work properly in various medical institutions, for different devices and different operators. This approach to analysis leads to the conclusion that it is impossible to obtain 100% sensitivity and specificity.

Extremely important advantages of infrared imaging, compared with other diagnostic techniques (using cameras operating in visible light or hyperspectral cameras), include ease of implementation of some image analysis and processing methods. This is due to the idea of measurement, namely calibrated pixels (of selected accuracy) and contrast for classical tests (patient warmer than the background). This simplifies segmentation and separation of the object from the background.

These issues and practical implementation of image analysis and processing methods are presented in the following chapters.

1.3   Nomenclature

Nomenclature adopted in this monograph refers to the orientation of the coordinate system and variable names.

All the graphs both in 2D and 3D are shown in the Cartesian coordinate system. Indexes relating to the coordinates of individual elements of the image matrix are determined in accordance with the nomenclature adopted in Matlab, FeeMat, Octave and Scilab. According to them, first the row position is given and then the column position. This is contrary to the Cartesian coordinate system (see Fig. 1.3).

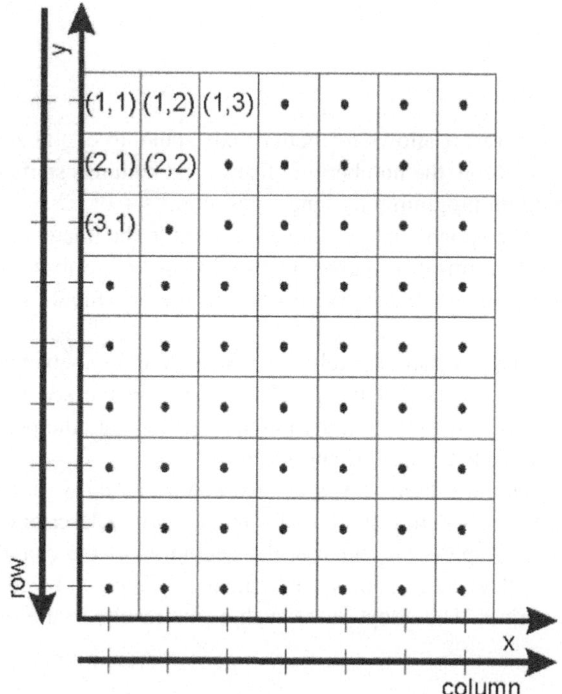

Fig. 1.3 Orientation of the cartesian coordinate system and the coordinate system adopted in accordance with Matlab

Fig. 1.4 Example of indicating the position of pixels for a 3D matrix sized $5 \times 4 \times 3$ pixels

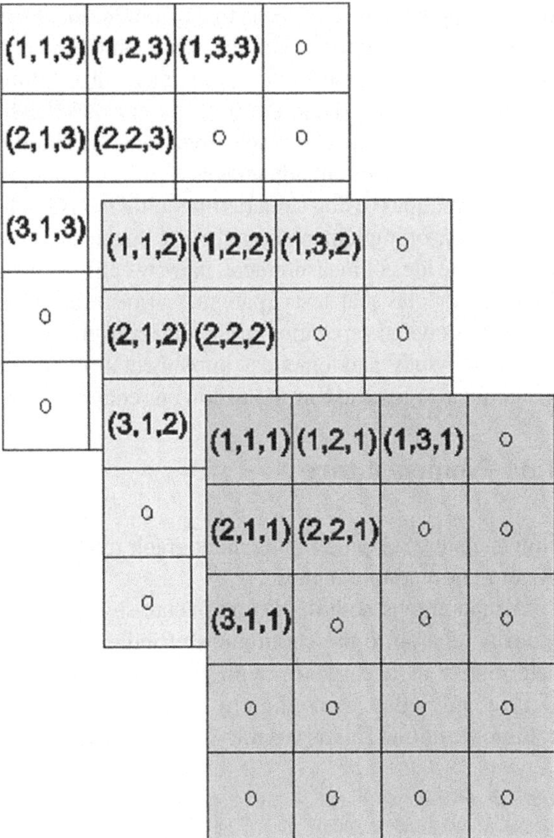

However, it allows for better understanding of the algorithms presented here. For this reason, the numbering of rows and columns starts with 1 and not 0—as in most modern programming languages, e.g. C++ or C#.

The variable names directly related to the image and representation of the image in the form of a matrix are labelled as "L". Subscripts are used to clarify what a given image relates to. In general, the L symbols without subscripts refer to any image.

The position of pixels in a row or column is given each time in parentheses, e.g. $L(2,3)$ means that the pixel is located in the second row and the third column. Subsequent figures in the parentheses indicate the position in the next dimensions. One such example is shown in Fig. 1.4.

The threshold values are marked with "p" with a subscript allowing to clarify what a given threshold relates to. A detailed description of the symbols used in this monograph is presented at the beginning of the chapter *List of selected symbols*.

All variable names are written in italics and not in bold even in the case of matrices. The adoption of such a rule results in the reference in any mathematical formula to the value of each pixel. All mathematical formulas also directly relate to integer, and not continuous, values. Unless specified otherwise, and the result of a specific operation or algorithm does not allow for the total numerical value, it is rounded to integer values.

The names of individual functions and scripts in Matlab are in italics. The presented source code is provided in a different font (Courier News) as opposed to the text of the book. The colours of the source code are directly linked to the colours visible in the debugger, i.e.:

green—comments,
red—strings,
blue—keywords,
black—text.

Chapter 2
Image Acquisition

Image acquisition is a typical process preceding image analysis and processing. The various stages of image acquisition should take into account: room size [51], infrared camera type [52, 53] (in particular its spatial resolution [54], temperature resolution and range of wavelengths), patient positioning and adequate preparation of the patient before examination (bearing in mind the restrictions described in the previous chapter) [55–59]. The distance of the infrared camera from the patient, artificial and natural sources of heat, and interference (noise) that can get into the measurement path play an important role during image acquisition.

Typical patient positioning relative to the infrared camera is shown in Fig. 2.1 [60, 61, 62].

The patient (Fig. 2.1) is typically standing 1.5–2 m away from the camera (depending on the focal length of the objective lens) [63, 64, 65]. The infrared camera is connected directly (via USB) to the computer equipped with data analysis software. The data obtained from the infrared camera can also be transferred using a data carrier, e.g. SD card/RAM. In this case, these are usually *.jpg*, *.bmp* files and the format tailored to Matlab, i.e. *.mat*, Further denoted as $L_{GRAY}(m,n)$. The application discussed in the monograph is tailored to the latter type of files (*.mat*). The computer used for data analysis was a PC with Microsoft Windows 7 Operating System Version 6.1 (Build 7601: Service Pack 1) and Intel® Core i7 3.6 GHz, 64 GB RAM. The Matlab version 7.11.0.584 (R2010b) and Image Processing Toolbox Version 7.1 (R2010b) were applied. The images were obtained retrospectively. They were acquired with various types of infrared cameras, starting with Agema 470 allowing for the acquisition of images with a spatial resolution of 140×140 pixels, through Agema 590 allowing for the acquisition of images with a spatial resolution of 320×240 pixels and ending with the FLIR A6750sc allowing for image acquisition with a resolution of 640×512 pixels. The temperature resolution mostly ranged from 0.1 to 0.05 [°C]. The length of the registered infrared radiation was in the range from 2 to 5 μm and further to 15 μm. This covers the range of peak, maximum emission for the human skin, which ranges from 9 to 12 μm-accurding to the analyses made by the American physiologist

© Springer International Publishing AG 2018
R. Koprowski, *Processing Medical Thermal Images*, Studies in Computational Intelligence 716, DOI 10.1007/978-3-319-61340-6_2

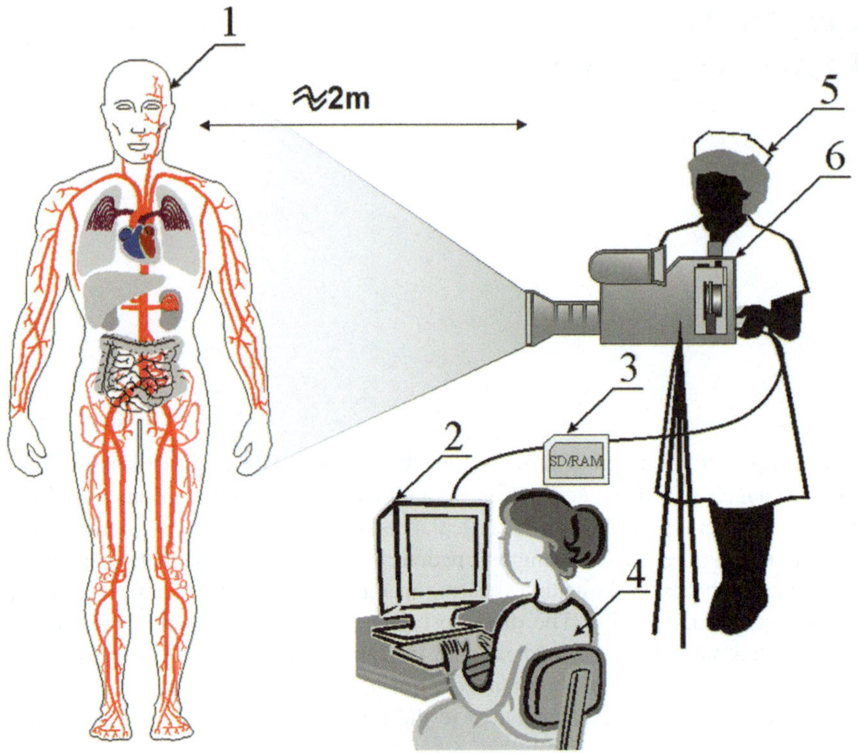

Fig. 2.1 Typical course of infrared image acquisition: *1* Patient; *2* Computer; *3* Memory card; *4* Computer operator; *5* Camera operator; *6* Infrared camera

J. D. Hardy (1934). Patients were aged 12–85 years. The patients suffered from different ailments, both dermatological and orthopaedic [66]. All patients gave voluntary consent to research. All measurements, which provided the images for analysis, were carried out in accordance with the Declaration of Helsinki. It should be emphasized here that no tests or measurements were performed on humans for the purposes of this monograph. The monograph is only tailored to the description of the methods and tools designed for analysis of the acquired thermal images. The next two chapters present image pre-processing and processing methods proposed by the author together with application and source code examples.

Chapter 3
Image Pre-processing

Image pre-processing is directly associated with the analysis of images acquired from an infrared camera and saved as a *.mat* file [67]. These images are further assigned to the variable $L_{GRAY}(m, n)$ where $m \in (1, M)$ and $n \in (1, N)$ as well as M and N are the number of rows and columns of the image matrix [68–70]. In general, this variable can be three-dimensional, i.e. $L_{GRAY}(m, n, i)$ where $i \in (1, I)$ and I is the total number of 2D images acquired in a single registration [71, 72].

These images were uploaded into the Matlab workspace using a graphical user interface (GUI). Presentation of this source code fragment will involve two *m-files*:

* *THV_GUI_MAIN*—GUI related m-file, which defines the position of various GUI elements (buttons, sliders, and others),
* *THV_GUI_sw(sw)*—function designed for reactions (callback) associated with the occurrence of an event—e.g. pressing a button defined in the function *THV_GUI_MAIN*.

Therefore, an image with the initially declared variable values will uploaded as follows (first fragments of the function *THV_GUI_MAIN* responsible for reading and saving the image):

```
%THV_GUI_MAIN graphical user interface
%    Copyright 2017 Robert Koprowski
%
%    Free only for readers book
global hObj LGRAY Lvar bin_var
scrsz = get(0,'ScreenSize');
LGRAY=rand(480,480); Lvar=LGRAY;
path='C:/THV';
cd(path);
hObj(1)=figure('Name','THV_GUI_MAIN',...
                'NumberTitle','off');
col=get(hObj(1),'Color')*1.1; col(col>1)=1;
```

© Springer International Publishing AG 2018
R. Koprowski, *Processing Medical Thermal Images*, Studies in Computational
Intelligence 716, DOI 10.1007/978-3-319-61340-6_3

```
hObj(2)=uicontrol('Style', 'pushbutton',...
                  'units','normalized',...
                  'FontUnits','normalized',...
                  'String', 'Open File(s)',...
                  'Position', [0.001 0.95 0.15 0.05],...
                  'Callback', 'THV_GUI_sw(1)',...
                  'BackgroundColor',col);
hObj(3)=uicontrol('Style', 'pushbutton',...
                  'units','normalized',...
                  'FontUnits','normalized',...
                  'String', 'Save File(s)',...
                  'Position', [0.001 0.9 0.15 0.05],...
                  'Callback', 'THV_GUI_sw(2)',...
                  'BackgroundColor',col);
```

The first lines preceded by '%' are a comment on the header of this *m-file*. Then, the global variables are declared:

- hObj—variable, vector, containing information relating to the holders of individual buttons,
- LGRAY—input image, uploaded, not altered during analysis and processing
- Lvar—image modified in subsequent stages of analysis,
- bin_var—values of binarization threshold.

Then an artificial image L_{GRAY} is created, of random pixel values of uniform distribution in the range from 0 to 1, i.e. LGRAY=rand(480,480); Lvar=LGRAY;. The next lines of the presented source code are path='C:/THV'; cd(path); positioning the path in which the images from the infrared camera were saved as a *.mat* file [73], [74], [75]. The next lines of the source code are window opening (hObj(1)=figure('Name'...), creating a palette of colours (col=get(hObj(1),'Color')*1.1; col(col>1)=1;) and double calling of the function uicontrol. The first calling of the function uicontrol is associated with the creation of the first button allowing for opening the *.mat* file. The second calling of the function uicontrol enables to create the second button for saving the image (saving the screenshot). Both the first and second calling of the function uicontrol was realized using similar parameters. These include parameters such as standardized units ('units','normalized'), text appearing on the button ('String', 'Save File(s)'), the position of the button ('Position', [0.001 0.9 0.15 0.05]), the aforementioned 'callback' and the background colour ('BackgroundColor',col). As a result of taking action (callback') in relation to the two buttons declared in the GUI, the function *THV_GUI_sw(sw)* is started. This function is activated with two arguments, for *sw=1* in the case of pressing the first button and *sw=2* in the case of the second button. The initial portion of the function *THV_GUI_sw(sw)* responsible for the action (for *sw=1 and sw=2*) of the first two buttons is as follows:

```matlab
function THV_GUI_sw(sw)
%THV_sw graphical user interface
%
%    Copyright 2017 Robert Koprowski
%
%    Free only for readers book
global hObj PathName FileName LGRAY excel_data Lvar bin_var

if sw==1 % read mat file
    [FileName,PathName] = uigetfile('*.mat','Select MAT file');
    if FileName~=0
        if strcmp(FileName(end-2:end),'mat')
            L=load([PathName,FileName]);
            FieldName=fieldnames(L);
            NumberLastImageStr=FieldName{end-3};
            nr_end=str2double(NumberLastImageStr(6:end));
            eval(['time=L.',FileName(1:end-4),'_DateTime;']);
            eval(['param=L.',FileName(1:end-4),'_ObjectParam;']);
            param=round(param*100)/100;
            eval(['LGRAY=L.Frame',mat2str(1),';']);
            set(hObj(19),'String', {['FileName=',FileName],...
                ['Date=',mat2str(time(1)),'\',
mat2str(time(2)),'\',mat2str(time(3))],...
                ['Time=',mat2str(time(4)),'\',mat2str(time(5)),'\',
mat2str(time(6)),'\.',mat2str(time(7))],...
                ['Emissivity=',mat2str(param(1))],...
                ['Distance=',mat2str(param(2))],...
                ['Reflected temp=',mat2str(param(3))],...
                ['Atmospheric temp=',mat2str(param(4))],...
                ['Relative hum=',mat2str(param(5))],...
                ['Trans.=',mat2str(param(6))]})
            set(hObj(15),'Min',1,'Max',size(LGRAY,1));
            set(hObj(16),'Min',1,'Max',size(LGRAY,2));
            bin_var=(round( (min(LGRAY(:)):0.1:max(LGRAY(:)))*10 )./10)';
            set(hObj(6),'String',{'None';num2str(bin_var)})
            excel_data=[];
            Lvar=LGRAY;
            THV_GUI_sw(4); % set slider
            THV_GUI_sw(3);
        else
            disp('This version read only *.mat files')
        end
    end
end
if sw==2 % write file
    Lgetframe=getframe(gcf);
    clk=clock;
    imwrite(Lgetframe.cdata,[[PathName,FileName],date,
mat2str(clk(3:end)),'.jpg'])
end
```

The first line defines the arguments of the function THV_GUI_sw(sw), and more precisely, the scalar value of the variable sw. The next lines refer to the comment on the modification date of function and copyright. Global variables relate to the following variables:

- hObj—handle objects,
- PathName—the name of the path for reading and saving files,
- FileName—the name of the read file and the name of the saved file,

- LGRAY—the read file,
- excel_data—the matrix of data to be saved in an Excel file *.xls*,
- Lvar—the image modified in subsequent transformations,
- bin_var—binarization threshold values selected for image temperature ranges.

For sw=1 the command string is tailored to reading *.mat* files. The first line `[FileName,PathName] = uigetfile('*.mat','Select MAT file');` enables to specify the file using the GUI implemented in Matlab. The results, two strings, are saved in the variables `FileName, PathName`. If `FileName~=0` and the last three letters of the file name are "*mat*" then the file is read using the function `load`. Next, the file attributes are saved in the variables *time* and *param*, and the image is saved in the variable *LGRAY*. In the next stage, a string of data obtained directly from the file is created, i.e.: *Emissivity, Distance, Reflected temp, 'Atmospheric temp, 'Relative hum, Trans* [76], [77], [78]. Next, the minimum and maximum values are set, attainable by moving the sliders (they are intended to show the profile of temperature changes and are discussed in the next chapter), i.e. `set(hObj(15),'Min',1,'Max',size(LGRAY,1));`. The data of individual binarization thresholds are saved in the variable *bin_var*. Then the variable *LGRAY* is copied to the variable *Lvar* and the message `'This version read only *.mat files'` is displayed if the last three letters in the file extension are not 'mat'.

For sw=2, the file being the screenshot in the same path as the read *.mat* file is saved. The name of the saved file consists of several components, i.e.: `PathName,FileName],date,mat2str(clk(3:end)),'.jpg'].` The first component is associated with the repetition of the name of the file being read. The second component is the date and the third one is the hour, minute and second. For example `n1.mat26-Nov-2016[26 15 20 47.58].jpg` means the read file `mat26` that was saved as a 'jpg' file on 26 November 2016 at 15:20:47.

The arrangement of these two buttons, namely *Open File(s)* and *Save File(s)*, and other menu items discussed in the following chapters is shown in Fig. 3.1.

The last element to discuss is the first image processing operation. This is median filtering. The input image L_{GRAY} is subjected to median filtering with a mask whose size is changed by the user in the range from 3×3 to 15×15 pixels every 2 pixels. The result is the image *Lvar*, and the source code is as follows (fragment of the function *THV_GUI_sw(sw)*):

```
hs=get(hObj(5),'Value');
if hs>1
    Lvar=medfilt2(Lvar,[hs*2-1, hs*2-1],'symmetric');
end
```

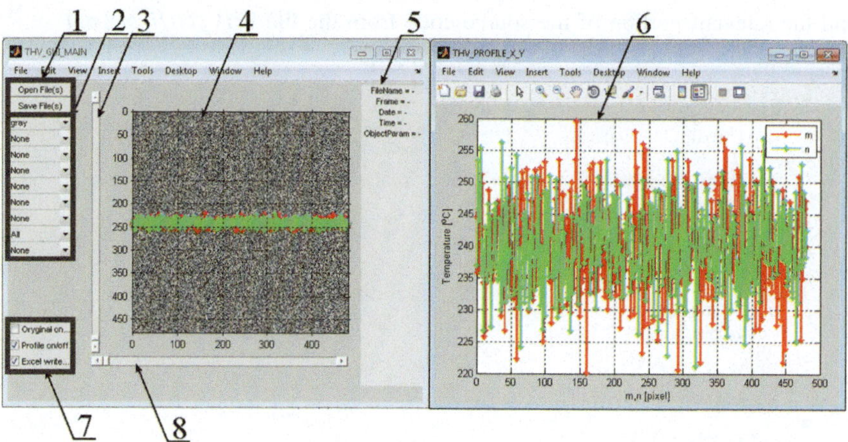

Fig. 3.1 Application main menu: *1* open and save buttons; *2* elements for selecting the next processing steps; *3* slider for the horizontal profile; *4* image L_{GRAY} or L_{var}; *5* information about the loaded file and acquisition parameters; *6* graph of temperature changes in the vertical and horizontal profile; *7* checkbox for viewing the image L_{GRAY} or L_{var}, profile viewing or saving to Excel; *8* slider for the vertical profile

where hObj(5) is the menu item defined in the function THV_GUI_MAIN i.e.:

```
hObj(5)=uicontrol('Style','popup',...
                  'units','normalized',...
                  'FontUnits','normalized',...
                  'String','None|Med 3x3|Med 5x5|Med 7x7|Med 9x9|
Med 11x11|Med 13x13|Med 15x15',...
                  'Position',[0.001 0.8 0.15 0.05],...
                  'Value',1,...
                  'Callback','THV_GUI_sw(3)',...
                  'BackgroundColor',col);
```

The selection of the appropriate option in 'popup' begins with the element 'None'. This is the first choice (*hs = 1*). For the next selection (of *hs* values), the mask size for the median filter changes according to the relation *hs·2-1*. In addition, of all the Matlab capabilities, the possibility of choosing the colour palette was introduced.

The relevant portions of the source code in the file *THV_GUI_MAIN* are as follows:

```
hObj(4)=uicontrol('Style', 'popup',...
                  'units','normalized',...
                  'FontUnits','normalized',...
                  'String','autumn|bone|colorcube|cool|copper|gray|
hot|hsv|jet|lines|pink|prism|spring|summer|white|winter',...
                  'Position',[0.001 0.85 0.15 0.05],...
                  'Value',6,...
                  'Callback','THV_GUI_sw(3)',...
                  'BackgroundColor',col);
```

and the relevant portion of the source code from the file *THV_GUI_sw(sw)*[1]

```
if sw==3
    Lvar=LGRAY;
    if get(hObj(4),'Value')==1
        colormap('autumn')
    end
    if get(hObj(4),'Value')==2
        colormap('bone')
    end
    if get(hObj(4),'Value')==3
        colormap('colorcube')
    end
    if get(hObj(4),'Value')==4
        colormap('cool')
    end
    if get(hObj(4),'Value')==5
        colormap('copper')
    end
    if get(hObj(4),'Value')==6
        colormap('gray')
    end
    if get(hObj(4),'Value')==7
        colormap('hot')
    end
    if get(hObj(4),'Value')==8
        colormap('hsv')
    end
    if get(hObj(4),'Value')==9
        colormap('jet')
    end
    if get(hObj(4),'Value')==10
        colormap('lines')
    end
    if get(hObj(4),'Value')==11
        colormap('pink')
    end
    if get(hObj(4),'Value')==12
        colormap('prism')
    end
    if get(hObj(4),'Value')==13
        colormap('spring')
    end
    if get(hObj(4),'Value')==14
        colormap('summer')
    end
    if get(hObj(4),'Value')==15
        colormap('white')
    end
    if get(hObj(4),'Value')==16
        colormap('winter')
    end
```

This stage ends image pre-processing operations.

---

[1]Similar fragments of the source code in which some parts of the source code are repeated are simplified later in the monograph.

Chapter 4
Image Processing

Image processing uses two types of images, L_{GRAY} and L_{VAR}. The described image processing is related to a sequence of transformations [79–95]:

- binarization [96],
- dilation [97],
- erosion [98],
- opening [99],
- closing [100],
- searching for the largest and smallest objects,
- hole filling.

The presented processing sequence results from previous author`s experience with biomedical image analysis. Therefore it is not an ad hoc method and is not tailored to a specific task [101, 102]. The overall objective of this processing stage is automatic detection of the object(s) in the scene. In the general case, it will enable to analyse the statistical values of the object temperature distribution, e.g. the patient`s feet or hands.

The first transformation is binarization. Similarly to image pre-processing, one matrix $LVAR$, defined in the source code, is modified and overwritten. For the correctness of writing mathematical formulas, substitute symbols with subscripts relating to individual transformations will be used below.

In the case of binarization, typical thresholding was used for the threshold p_r selected manually by the operator in the GUI, i.e. image L_{BIN} after binarization is equal to:

$$L_{BIN}(m, n) = \begin{cases} 1 & if & L_{GRAY}(m, n) > p_r \\ 0 & other \end{cases} \tag{4.1}$$

© Springer International Publishing AG 2018

R. Koprowski, *Processing Medical Thermal Images*, Studies in Computational Intelligence 716, DOI 10.1007/978-3-319-61340-6_4

The corresponding fragment in the m-file *THV_GUI_sw(sw)* is presented below:

```
pr=get(hObj(6),'Value');
if pr>1
    Lvar=Lvar>(bin_var(pr-1));
end
```

where the relevant fragment was located in the m-file *THV_GUI_MAIN*:

```
bin_var=(0:0.1:1)';
hObj(6)=uicontrol('Style','popup',...
                  'units','normalized',...
                  'FontUnits','normalized',...
                  'String',{'None';num2str(bin_var)},...
                  'Position',[0.001 0.75 0.15 0.05],...
                  'Value',1,...
                  'Callback','THV_GUI_sw(3)',...
                  'BackgroundColor',col);
```

The values of the variable `bin_var` are selected by choosing the subsequent elements from `popup` (`hObj(6)`). It should be noted here that in the variable `bin_var` the values range from 0 to 1 every 0.1 when running the script for the first time (the values of random matrices L_{GRAY}—see Fig. 3.1). After uploading any file *.mat*, the range of acceptable binarization threshold values is changed. It is then from the maximum to minimum value every 0.1 [°C], i.e.: $bin\_var \in (\min(L_{GRAY}), \max(L_{GRAY}))$, and the corresponding source code fragment is as follows:

```
bin_var=(round( (min(LGRAY(:)):0.1:max(LGRAY(:)))*10)./10)';
set(hObj(6),'String',{'None';num2str(bin_var)})
```

Rounding seen in the above fragment of the source code results from the need to obtain the accuracy of one decimal place (0.1 [°C]).

The second transformation is dilation. The idea of dilation involves using the function imdilate implemented in Matlab. The following mathematical relation is used:

$$L_{DIL}(m,n) = \max_{m_{SED},n_{SED} \in SED} \left(L_{GRAY}(m+m_{SED}, n+n_{SED}) \right) \tag{4.2}$$

where:

L_{DIL} —image resulting from dilation,
SED —structural element (*SE*) with row and column coordinates (m_{SED}, n_{SED})

In simple terms:

$$L_{DIL} = L_{GRAY} \oplus SED \tag{4.3}$$

and the corresponding source code:

```
SE=get(hObj(7),'Value');
    if SE>1
        Lvar=imdilate(Lvar,ones([SE*2-1, SE*2-1]));
    end
```

and

```
hObj(7)=uicontrol('Style', 'popup',...
                  'units','normalized',...
                  'FontUnits','normalized',...
                  'String','None|Dil 3x3|Dil 5x5|Dil 7x7|Dil 9x9|
Dil 11x11|Dil 13x13|Dil 15x15',...
                  'Position',[0.001 0.7 0.15 0.05],...
                  'Value',1,...
                  'Callback','THV_GUI_sw(3)',...
                  'BackgroundColor',col);
```

The range of changes in the size of the structural element is limited from 3×3 to 15×15 pixels.

The third transformation is erosion. In a manner analogous to dilation, it uses the implemented function `imerode`. The following formula is used in this case:

$$L_{ERO}(m,n) = \min_{m_{SEE},n_{SEE} \in SEE} (L_{GRAY}(m+m_{SEE}, n+n_{SEE})) \tag{4.4}$$

where:

L_{ERO} —image resulting from erosion,
SEE —structural element (SE) with row and column coordinates (m_{SEE}, n_{SEE})

In simple terms:

$$L_{ERO} = L_{GRAY} \ominus SEE \tag{4.5}$$

The corresponding source code fragment is as follows:

```
SE=get(hObj(8),'Value');
if SE>1
   Lvar=imerode(Lvar,ones([SE*2-1, SE*2-1]));
end
```

and

```
hObj(8)=uicontrol('Style', 'popup',...
                  'units','normalized',...
                  'FontUnits','normalized',...
                  'String', 'None|Ero 3x3|Ero 5x5|Ero 7x7|Ero 9x9|
Ero 11x11|Ero 13x13|Ero 15x15|',...
                  'Position',[0.001 0.65 0.15 0.05],...
                  'Value',1,...
                  'Callback','THV_GUI_sw(3)',...
                  'BackgroundColor',col);
hObj(9)=uicontrol('Style','popup',...
```

The fourth and fifth transformations are opening and closing. Similarly to the previously discussed transformations (erosion and dilation), they are based on the functions `imopen` and `imclose`, intended for morphological opening and closing respectively. These operations are associated with the next menu elements and the following source code fragments:

```
hObj(9)=uicontrol('Style','popup',...
                  'units','normalized',...
                  'FontUnits','normalized',...
                  'String','None|Ope 3x3|Ope 5x5|Ope 7x7|Ope 9x9|
Ope 11x11|Ope 13x13|Ope 15x15|',...
                  'Position',[0.001 0.6 0.15 0.05],...
                  'Value',1,...
                  'Callback','THV_GUI_sw(3)',...
                  'BackgroundColor',col);
hObj(10)=uicontrol('Style','popup',...
                  'units','normalized',...
                  'FontUnits','normalized',...
                  'String','None|Clo 3x3|Clo 5x5|Clo 7x7|Clo 9x9|
Clo 11x11|Clo 13x13|Clo 15x15|',...
                  'Position',[0.001 0.55 0.15 0.05],...
                  'Value',1,...
                  'Callback','THV_GUI_sw(3)',...
                  'BackgroundColor',col);
```

and

```
SE=get(hObj(9),'Value');
if SE>1
   Lvar=imopen(Lvar,ones([SE*2-1, SE*2-1]));
end
SE=get(hObj(10),'Value');
if SE>1
   Lvar=imclose(Lvar,ones([SE*2-1, SE*2-1]));
end
```

The formal record of morphological opening (L_{OPE}) and closing (L_{CLO}) is as follows:

$$L_{OPE} = (L_{GRAY} \ominus SEE) \oplus SED \tag{4.6}$$

$$L_{CLO} = (L_{GRAY} \oplus SED) \ominus SEE \tag{4.7}$$

The sixth transformation is searching for the largest or smallest object. This operation is directly related to the labelling function, i.e. `bwlabel`. Implementation of this function into binary images enables to control individual objects, in terms of both surface area and the selection of the largest or smallest object. Based on the author's past experience, the following options have been selected:

- selection of one largest object,
- selection of two largest objects,
- selection of three largest objects,
- selection of the smallest object.

The corresponding source code is as follows:

```
if get(hObj(11),'Value')~=1
    if sum(sum(Lvar))>0
    if sum(sum((Lvar~=1)&(Lvar~=0)))==0
Llab=bwlabel(Lvar); pam_pow=[];
        for il=1:max(Llab(:))
            pam_pow(il,1:2)=[sum(sum(Llab==il)) il];
        end
        pam_pow_1=sortrows(pam_pow,-1);
        if (size(pam_pow_1,1)>=1) && (get(hObj(11),'Value')==2)
            Lvar=Llab==pam_pow_1(1,2);
        end
        if (size(pam_pow_1,1)>=2) && (get(hObj(11),'Value')==3)
            Lvar=(Llab==pam_pow_1(1,2))|(Llab==pam_pow_1(2,2));
        end
        if (size(pam_pow_1,1)>=3) && (get(hObj(11),'Value')==4)
            Lvar=(Llab==pam_pow_1(1,2))|(Llab==pam_pow_1(2,2))|
(Llab==pam_pow_1(3,2));
        end
        if (size(pam_pow_1,1)>=1) && (get(hObj(11),'Value')==5)
            Lvar=Llab==pam_pow_1(end,2);
        end
    else
        warndlg('It''s NOT binary image')
    end
    end
end
```

and

```
hObj(11)=uicontrol('Style','popup',...
                'units','normalized',...
                'FontUnits','normalized',...
                'String','All|1Max Obj|2Max Obj|3Max Obj|Min Obj|',...
                'Position',[0.001 0.5 0.15 0.05],...
                'Value',1,...
                'Callback','THV_GUI_sw(3)',...
                'BackgroundColor',col);
```

The following is the key fragment of the source code:

```
Llab=bwlabel(Lvar); pam_pow=[];
        for il=1:max(Llab(:))
            pam_pow(il,1:2)=[sum(sum(Llab==il)) il];
        end
        pam_pow_1=sortrows(pam_pow,-1);
```

in which subsequent *il* images are analysed. Next the results are saved in the variable `pam_pow` and then sorted (relative to the surface area). If the input image is not binary, the warning `'It''s NOT binary image'` appears.

The seventh transformation is hole filling. It involves filling holes of all the objects in the binary input image [103, 104]. The corresponding source code fragments are given below:

```
hObj(12)=uicontrol('Style', 'popup',...
                   'units','normalized',...
                   'FontUnits','normalized',...
                   'String','None|Fill',...
                   'Position',[0.001 0.45 0.15 0.05],...
                   'Value',1,...
                   'Callback','THV_GUI_sw(3)',...
                   'BackgroundColor',col);
```

and

```
if get(hObj(12),'Value')==2
    Lvar=bwfill(Lvar,'holes');
end
```

The GUI intended for selecting the options of the discussed transformations is shown in Fig. 4.1.

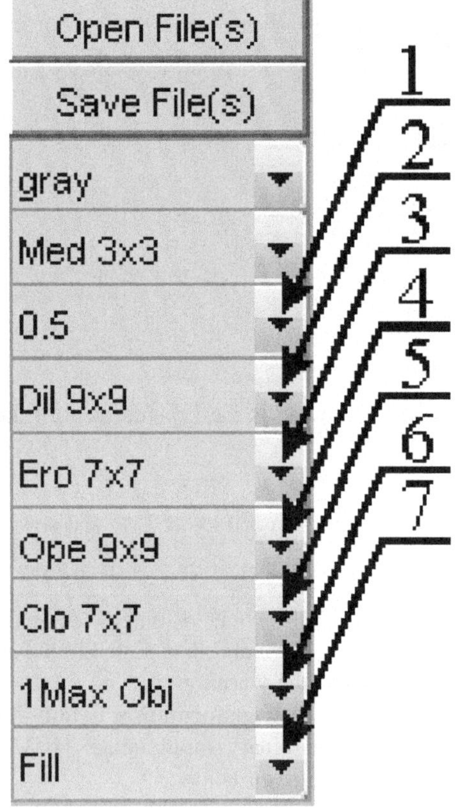

Fig. 4.1 Menu for selecting the next processing steps (sample set values); *1* binarization, *2* dilation, *3* erosion, *4* opening, *5* closing, *6* searching for the largest and smallest objects, *7* hole filling

Additional functions

The following two functions are extremely useful during image processing:

- analysis of temperature changes in the vertical and horizontal profile [105, 106, 107] line and
- basic calculations and saving statistical information on the temperature in the area of the object.

Analysis of temperature changes is associated with visualization of temperature changes on the graph (Fig. 3.1,6) for each horizontal and vertical pixel. The line position is changed manually by means of two sliders arranged directly next to the displayed image (Fig. 3.1, 3, 8).

The corresponding source code allowing for profile line mixing, both vertically and horizontally (green and red respectively), is as follows:

```
if sw==4 % slider
    ww=get(hObj(15),'Max')-round(get(hObj(15),'Value'));
    kk=round(get(hObj(16),'Value'));
    set(hObj(17),'XData',[kk kk],'YData',[1 size(Lvar,1)]);
    set(hObj(18),'YData',[ww ww],'XData',[1 size(Lvar,2)]);
    set(hObj(25),'YData',Lvar(:,kk),'XData',[1:size(Lvar,1)]);
    set(hObj(26),'YData',Lvar(ww,:),'XData',[1:size(Lvar,2)]);
end
```

and

```
hObj(15)=uicontrol('Style', 'slider',...
                   'units','normalized',....
                   'FontUnits','normalized',...
                   'BackgroundColor',col,...
                   'Position', [0.2 0.15 0.02 0.82],...
                   'Min',1,'Max',480,'Value',1,...
                   'SliderStep',[1/100 1/100],...
                   'Callback','THV_GUI_sw(4)');
hObj(16)=uicontrol('Style','slider',...
                   'units','normalized',....
                   'FontUnits','normalized',...
                   'BackgroundColor',col,...
                   'Position',[0.2 0.11 0.62 0.03],...
                   'Min',1,'Max',480,'Value',1,...
                   'SliderStep',[1/100 1/100],...
                   'Callback','THV_GUI_sw(4)');
hObj(17)=plot(sum(LGRAY,1),'-r*','LineWidth',2,'Parent',
hObj(13)); hold on; grid on
hObj(18)=plot(sum(LGRAY,2),'-g*','LineWidth',2,'Parent',hObj(13));
hObj(23)=figure('Name','THV_PROFILE_X_Y',...
                'NumberTitle','off');
hObj(24)=axes('Parent',hObj(23),...
              'units','normalized',...
              'Position', [0.1 0.1 0.85 0.85]);
hObj(25)=plot(sum(LGRAY,1),'-r*','LineWidth',2,'Parent',
hObj(24)); hold on; grid on
hObj(26)=plot(sum(LGRAY,2),'-g*','LineWidth',2,'Parent',hObj(24));
                xlabel('m,n [pixel]')
                ylabel('Temperature [^oC]')
                legend('m','n')
```

Fig. 4.2 Block diagram of
the proposed transformations.
Different background colours
indicate the three stages:
image acquisition—green,
image pre-processing—blue
and image processing—red

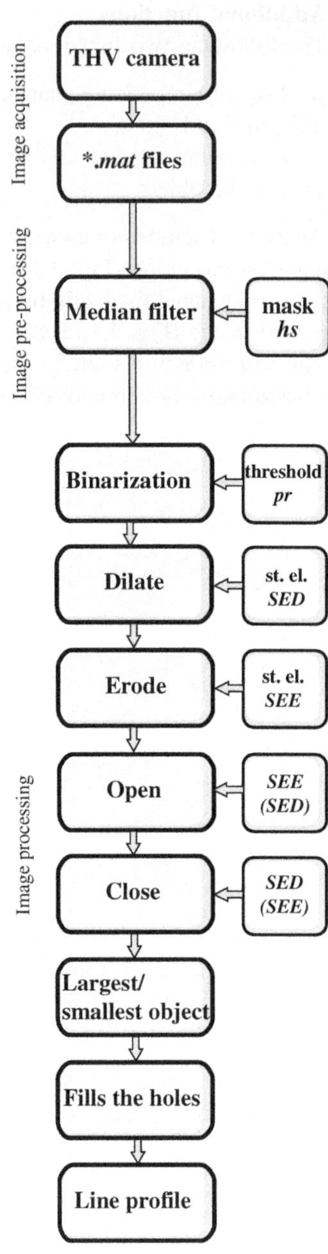

Another additional feature is the ability to automatically save the results to an
Excel file. This is done in the following source code fragment:

```
hObj(22)=uicontrol('Style','checkbox',...
                   'units','normalized',...
                   'FontUnits','normalized',...
                   'String','Excel write on/off',...
                   'Position',[0.001 0.10 0.15 0.05],...
                   'Value',1,...
                   'Callback','THV_GUI_sw(6)',...
                   'BackgroundColor',col);
```

and

```
if sw==6
    if get(hObj(22),'Value')==1
    if ~isempty(excel_data)
        clk=clock;
        xlswrite([[PathName,FileName],date,'.xls'],excel_data);
    end
    end
end
```

Data derived from the data vector `excel_data` are saved to Excel. The vector
is then supplemented with new data at every processing stage. The prepared and
discussed source code fragment only enables to save two data - minimum and
maximum temperature of the object, i.e.:

```
excel_data=double([min(Lvar(:)) max(Lvar(:))]);
```

Figure 4.2 shows a block diagram of image acquisition and processing that have
been completed so far.

At this point, I encourage readers to introduce their own modifications to the
presented algorithm, for example, add new features, capabilities or even increase
the number of data saved to Excel. The source code is a set of basic data and is
stored separately in the appendix to this monograph.

In the following chapters, the presented algorithm and capabilities are modified,
overwritten and extended. In particular, the presented algorithm is extended to
include analysis of thermal image sequences.

Chapter 5
Examples of Tailoring the Algorithm

Currently, no algorithm is versatile enough to allow for automated and repeatable measurement of all parameters for any images. There are two possible directions of designing applications intended for automated image analysis. The first involves implementation of all known and new methods of image analysis and processing. The second direction involves tailoring applications to measurement automation for a single type of task/analysis. Tailoring an algorithm to a particular application has great advantages such as full automation of measurements of hundreds of thousands of images without operator intervention. They provide quantitative and reproducible diagnostic information. Their drawback is, in addition to the need to tailor them, the possibility of over-fitting of the device and operator to the data. Consequently, their use for the same purpose, e.g. intraocular pressure (IOP) measurement [108] or the calculation of the surface area of dermatoses in dermatology [109], is the cause of significant differences in the results. On the other hand, it is extremely difficult to carry out tests of a tailored algorithm based on images from different parts of the world taken by different operators for patients of different races. Repositories of medical images placed in various locations on the Internet still do not cover the full spectrum of possibilities and diseases. Thus, one possibility is to present a tailored algorithm designed for the measurement of a single disease entity. The reader, by testing and comparing their own data, will be able to make the necessary adjustments so that the algorithm will be fully automatic and the results reproducible. Among the various applications of infrared imaging in medicine and tailored algorithms, the following were selected:

- automatic analysis of temperature on the patient's back [110–112],
- quantitative assessment of the correctness of performing dermatological laser treatments [113, 114],
- assessment of thermal comfort in newborns [115–117],
- assessment of headaches on the basis of thermal images of the face [118–120].

A detailed description of the implemented methods is presented in the following chapters.

© Springer International Publishing AG 2018

R. Koprowski, *Processing Medical Thermal Images*, Studies in Computational Intelligence 716, DOI 10.1007/978-3-319-61340-6_5

These methods are accessible directly from the main window of the presented application (Fig. 4.1) in the form of buttons located in the upper part of the window. This requires modification of the existing source code (files *THV_GUI_MAIN* and *THV_GUI_sw(sw)*). Accordingly, the new modified files are placed in a separate container attached to this book.

5.1   Analysis of Temperature Distribution on the Patient's Back

One of the applications of infrared imaging is the assessment of the asymmetry degree of temperature distribution on the patient's back. In terms of diagnosis this is valuable information that allows for early diagnosis of skin infections, some diseases of internal organs and scoliosis. Issues concerning the analysis of asymmetry in temperature distribution on the patient's back or other areas of their body is primarily associated with the selection of a gold standard. In general, it is a control group of healthy patients, other medical evidence or results of theoretical analyses. In the case of infrared imaging, it is mostly a group of healthy individuals, because the thermoregulation process is multifactorial [121, 122]. Thus, designing an algorithm is simplified to creating an automated and repeatable method of analysis of the (healthy and ill) subject's back. The main problem here is inter-subject variability. In this case, namely the patient's back, the main problem is its variable size. Accordingly, a method was proposed that enables to standardize the variable size of the patient's back through the analysis of data in a grid with a fixed number of squares. In summary, the analysis of temperature distribution on the patient's back allows for:

- quantitative assessment of temperature distribution maintaining repeatability of measurements for patients of different physical constitution,
- determination of the outline of the spine, in particular its potential curvature,
- comparison of temperature distribution correctness with the standard,
- assessment of the statistical parameters of temperature distribution such as gray-level co-occurrence matrix (GLCM) [123–127], square-tree decomposition [128, 129] or morphological features [130–132].

The initial stages of analysis of temperature distribution on the torso are directly related to the detection of the position of characteristic points $l_1, l_2, l_3, l_4, l_5, l_6, l_7, l_8$ shown in Fig. 5.1.

The position of the characteristic points (according to Fig. 5.1) is as follows (around organs):

- l_2, l_3—neck,
- l_1, l_4—shoulders,
- l_5, l_8—armpits,
- l_6, l_7—hips.

Fig. 5.1 Explanatory figure showing the distribution of characteristic points on the patient's back together with a grid sized 12×8 pixels

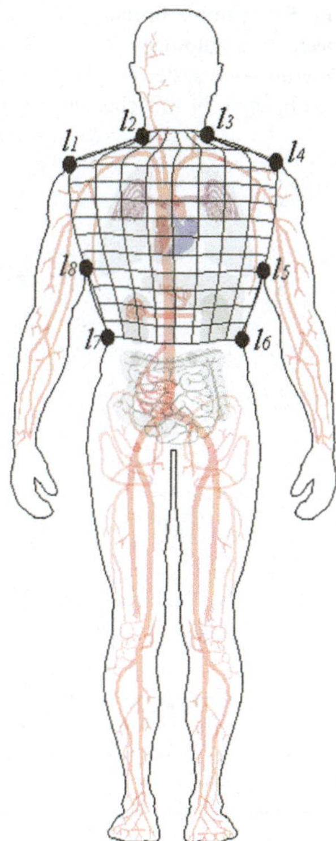

The analysis is carried out based on the binary image L_{BI2} that contains an object of the maximum surface area visible in the image L_{BIN} (see the paragraph "The sixth transformation is …" in the previous chapter). The objective of these transformations is shown in Fig. 5.2. This is an automatic and repeatable temperature analysis of the back.

In the previous paragraph, according to the adopted assumption, subsequent transformations modified and overwrote the contents of the image L_{VAR}. In this and the following subchapters, the order of indexing is maintained (starting from the binary image L_{BI2}).

The first stage involves designation of the position of points (pixels with the value "1") for which there is only one object horizontally—see Fig. 5.1.

Data vectors L_S and L_{SV} (Fig. 5.3) necessary for further analysis are calculated as follows:

$$L_S(m) = \max_{n \in (1,N)} L_{SF}(m,n) \tag{5.1}$$

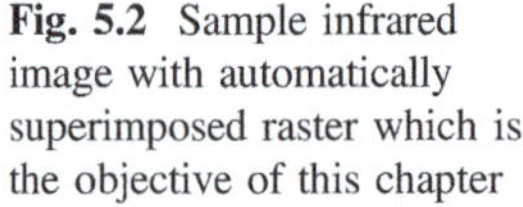

Fig. 5.2 Sample infrared image with automatically superimposed raster which is the objective of this chapter

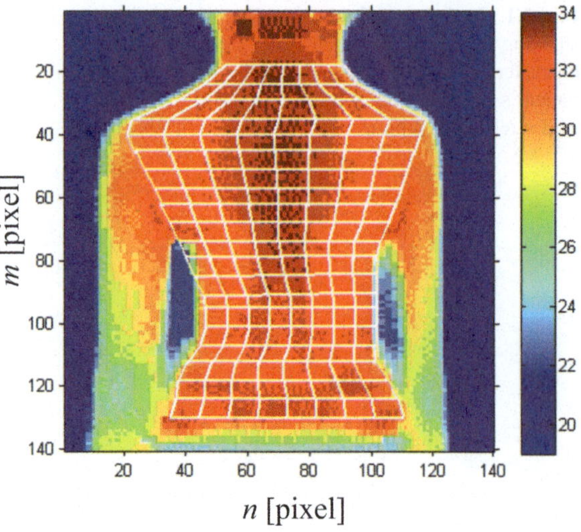

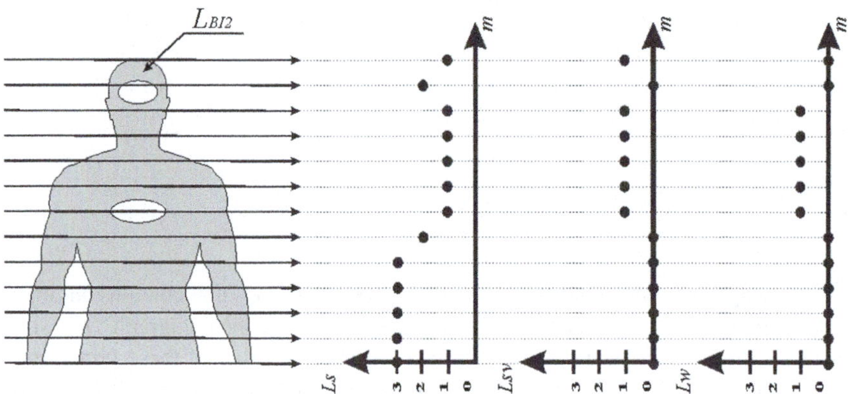

Fig. 5.3 Explanatory figure showing some selected stages of analysis of the largest continuous object relative to the y-axis

where:

$$L_{SF}(m,n) = \begin{cases} \max\limits_{m\in(1,m-1)}\left(L_{SF}(m,n)\right)+1 & if \quad L_{BI2}(m,n) - L_{BI2}(m-1,n) = 1 \\[2ex] \max\limits_{m\in(1,m-1)}\left(L_{SF}(m,n)\right) & if \quad \begin{aligned}&(L_{BI2}(m,n) - L_{BI2}(m-1,n) = 0)\\ &\wedge(L_{BI2}(m,n) = 1)\end{aligned} \\[2ex] 0 & if \quad \begin{aligned}&(L_{BI2}(m,n) - L_{BI2}(m-1,n) = 0)\\ &\wedge(L_{BI2}(m,n) = 0)\end{aligned} \end{cases} \tag{5.2}$$

for $m \in (2,M)$.

The values of L_S and L_{SV} are calculated according to the following source code:

```
Lsv=zeros([1 size(LBI2,1)]);
   for i=1:size(LBI2,1)
      Ls=bwlabel(LBI2(i,:));
      if max(Ls)==1
         Lsv(i)=1;
      end
   end
```

L_{SV} values are therefore designated as:

$$L_{SV}(m) = \begin{cases} 1 & if \quad L_S(m) = 1 \\ 0 & other \end{cases} \tag{5.3}$$

Next, the largest object is designated, vector L_W, i.e.:

$$L_W(m) = \begin{cases} 1 & if \quad L_{WF}(m) = \arg\max_{k\in(1,K)} L_{WFK}(k) \\ 0 & other \end{cases} \tag{5.4}$$

where:

$$L_{WFK}(k) = \begin{cases} \sum_{m=1}^{M} L_{WF}(m) & if \quad L_{WF}(m) = k \\ 0 & other \end{cases} \tag{5.5}$$

$$L_{WF}(m) = \begin{cases} \max_{m\in(1,m-1)}(L_{WF}(m))+1 & if \quad L_{SV}(m) - L_{SV}(m-1) = 1 \\ \max_{m\in(1,m-1)}(L_{WF}(m)) & if \quad \begin{matrix}(L_{SV}(m) - L_{SV}(m-1) = 0)\\ \wedge(L_{SV}(m) = 1)\end{matrix} \\ 0 & if \quad \begin{matrix}(L_{SV}(m) - L_{SV}(m-1) = 0)\\ \wedge(L_{SV}(m) = 0)\end{matrix} \end{cases} \tag{5.6}$$

for $m \in (2,M)$.

The corresponding source code fragment is as follows:

```
LWF=bwlabel(Lsv);
   po=0;
   for k=1:max(LWF)
    [oo,pp]=find(LWF==k);
      if po<length(pp)
         po=length(pp);
         ppo=k;
      end
   end
   LW=(LWF==ppo);
```

The next step involves analysis of the image L_{BI4} defined as:

$$L_{BI4}(m,n) = L_{BI3}(m,n) \cdot L_{BI2}(m,n) \tag{5.7}$$

where:

$$L_{BI3}(m,n) = \begin{cases} 1 & if \quad L_W(m) = 1 \\ 0 & other \end{cases} \tag{5.8}$$

In the second stage, the position of the neck is detected. This detection involves calculating the derivative of the total value of the elements in each row of the image L_{BI4}, i.e.:

$$L_{BG4}(m) = L_{BS4}(m) - L_{BS4}(m+1) \tag{5.9}$$

where:

$$L_{BS4}(m) = \sum_{n=1}^{N} L_{BI4}(m,n) \tag{5.10}$$

and $m \in (1, M-1)$.

Vector L_{BG4}[1] includes information associated with the change in the sum of the horizontal pixel values. Sample results are shown in Figs. 5.4 and 5.5.

The corresponding source code fragment is as follows:

```
LBI3=(LW'*ones([1 size(LBI2,2)]));
   imshow(LBI3,[],'notruesize');
   LBI4=LBI3.*LBI2;
   LBS4=sum(LBI4');
   figure; plot(LBS4,'-b*');
   ylabel('L_{BS4}(m)','FontSize',15,'FontAngle','Italic')
   xlabel('m [pixel]','FontSize',15,'FontAngle','Italic');
   grid on
   LBG4=gradient(LBS4);
   LBG5=LBG4.*(imerode(LBS4>0,ones(1,3)));
   figure('Color',[1 1 1])
   plot(LBG5,'k');
   xlabel('m [pixel]','FontSize',15,'FontAngle','Italic')
   ylabel('L_{BG5}(m)','FontSize',15,'FontAngle','Italic')
   grid on
   [~,n]=find(LBG4==max(LBG5));
   [~,n2]=find(fliplr(LBG5(1:n(1)))<=0);
   if isempty(n2); n2=n(1)-1; end
   n=n(1)-n2(1)+1;
   [~,pp]=find(LBI4(n(1),:)==1);
   figure('Color',[1 1 1]);
   imshow(LGRAY,[],'notruesize');
   colormap('jet');
```

---

[1]In order to standardize the subscripts and provide uniform record, the product (matrix, vector or scalar) resulting from analysis assumes the same numbering in the subscripts. For that reason, the existing L_{BG4} does not necessarily mean that there is L_{BG3} or L_{BG2}.

```
colorbar;
l213=[pp(1) pp(length(pp)) n(1)];
hold on
text(l213(1),l213(3),'+','Color',
[1 1 1],'FontSize',28,'HorizontalAlignment','center');
text(l213(2),l213(3),'+','Color',
[1 1 1],'FontSize',28,'HorizontalAlignment','center');
```

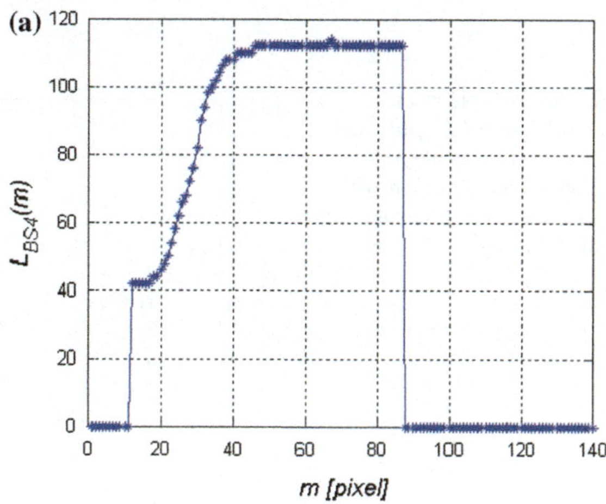

Fig. 5.4 Sample results of analysis of L_{BS4}: **a** graph of L_{BS4} as a function of subsequent rows; **b** image L_{BI2} on whose basis the graph in (**a**) was obtained

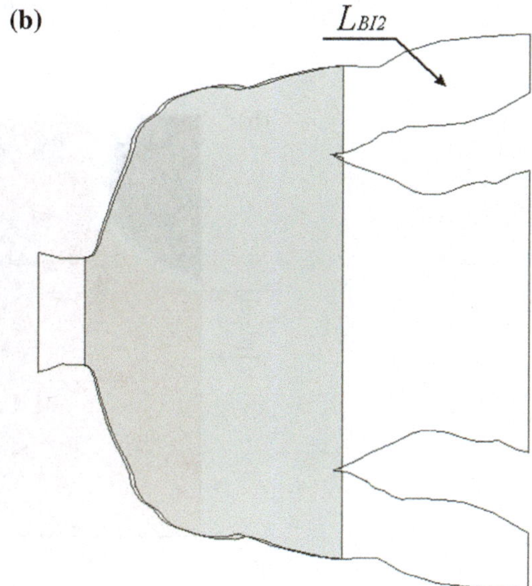

In the third step, shoulder appendages are located—points l_1, l_4. For this purpose, the image L_{BI4} is divided into two parts. The dividing line n_{med} is the mean of the number of pixels of the object (object surface) on the right and left side of the division, i.e.:

```
n_=zeros([1 size(LBI4,1)]);
for i=1:size(LBI4,1)
    [pp,oo]=find(LBI4(i,:)==1);
    if ~isempty(oo)
        n_(i)=round((oo(length(oo))-oo(1))/2+oo(1));
    end
end
n_(n_==0)=[];
n_med=round(mean(n_(:)));
```

Fig. 5.5 Sample results: **a** for the vector $L_{BG5}(m)$ and **b** the position of the neck—points l_2 and l_3 marked with *white* crosses in a sample infrared image

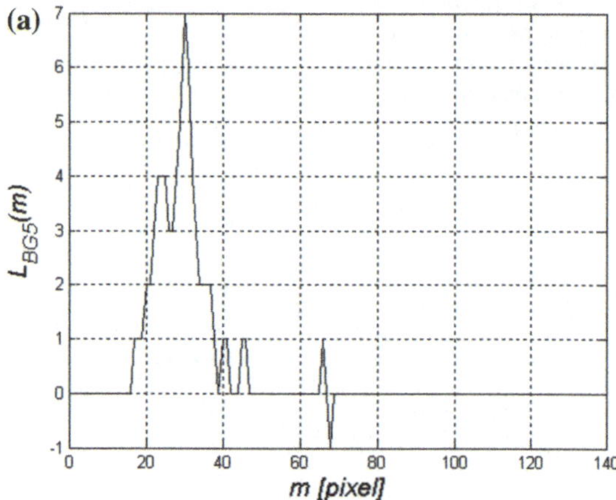

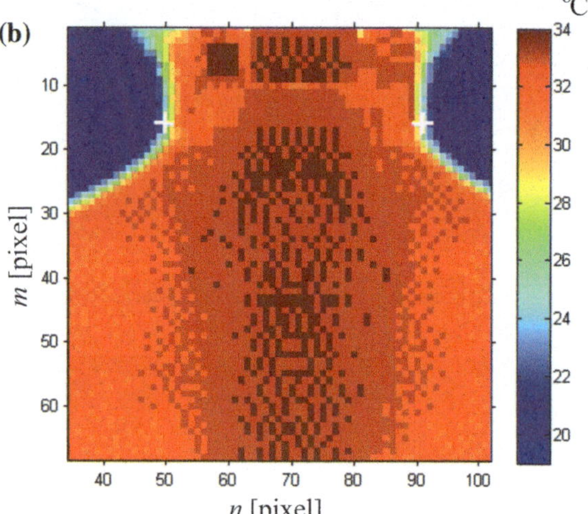

After dividing the object into images L_{BI5L} and L_{BI5P}:

$$L_{BI5L} = L_{BI5}(m, n) \tag{5.11}$$

for $m \in (1,M)$ and $n \in (1,n_{med})$. In addition

$$L_{BI5P} = L_{BI5}(m, n) \tag{5.12}$$

for $m \in (1,M)$ and $n \in (n_{med},N)$.

Radon operation is performed on both halves of the image L_{BI5} for the results obtained, for example:

$$R_{ad}(n', \alpha_L) = \sum_{m'} L_{BI5L}(n'\cos(\alpha_L) - m'\sin(\alpha_L), n'\sin(\alpha_L) - m'\cos(\alpha_L)) \tag{5.13}$$

where:

$$\begin{bmatrix} n' \\ m' \end{bmatrix} = \begin{bmatrix} \cos(\alpha_L) & \sin(\alpha_L) \\ -\sin(\alpha_L) & \cos(\alpha_L) \end{bmatrix} \begin{bmatrix} n \\ m \end{bmatrix} \tag{5.14}$$

The division into two images and Radon operation are completed as shown in Figs. 5.6 and 5.7.

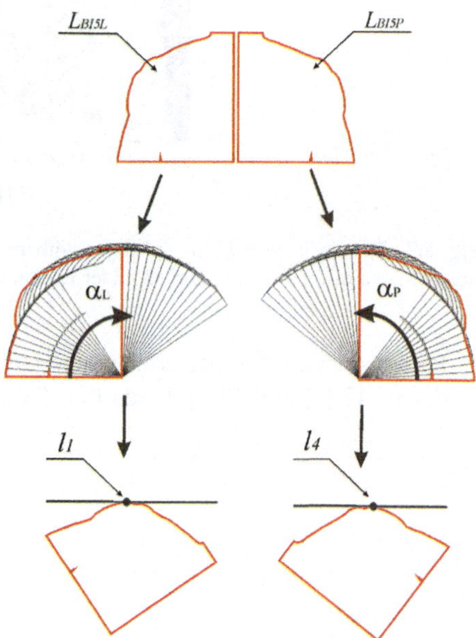

Fig. 5.6 Schematic diagram of the method for calculating the position of points l_1 and l_4 on the halves of the image L_{BI5}, i.e. L_{BI5L} and L_{BI5P}

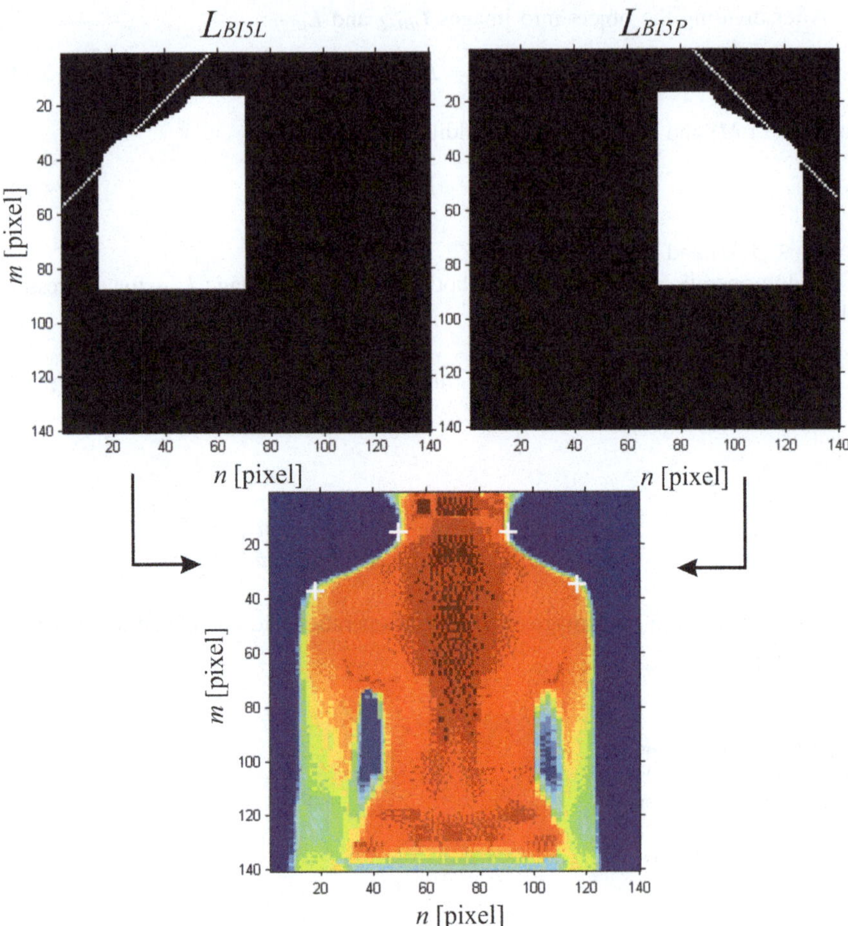

Fig. 5.7 Sample images L_{BI5L} and L_{BI5P} with marked tangents and an infrared image with the marked position of the neck and shoulder appendages

Radon operation is performed separately for images L_{BI5L} and L_{BI5P}, for fixed angles of 45 [°] and 135 [°] (see Fig. 5.6), i.e.:

```
LBI5L=LBI4; LBI5L(1:l213(3),:)=0;
LBI5L(:,1:n_med(:))=0;

figure('Color',[1 1 1]);
imshow(LBI5L,[],'notruesize');

figure
rad=radon(LBI5L,45);

[pp,~]=find(rad>0);
dlk=round((length(rad)-pp(length(pp)))/sqrt(2))+1;
LBI6=zeros(size(LBI5L));
```

```
   for i=1:(2*dlk-1)
       LBI6((dlk*2-i),((140-dlk*2)+(2*dlk-i)))=1;
   end
figure; imshow(LBI5L|LBI6)
LBI7=bwlabel(LBI5L.*LBI6);
LBI8=(LBI7==max(LBI7(:)));
[oo,pp]=find(LBI8==1);
14=[mean(pp), mean(oo)];
  LBI5P=LBI4; LBI5P(1:1213(3),:)=0;
  LBI5P(:,n_med:end)=0;
  rad=radon(LBI5P,135);
  [pp,~]=find(rad>0);
  dlk=round(((length(rad)-pp(length(pp)))/sqrt(2))+1;
  LBI6=zeros(size(LBI5P));
     for i=1:(2*dlk-1)
         LBI6((dlk*2-i),i)=1;
     end
  figure; imshow(LBI5P|LBI6)
  LBI7=bwlabel(LBI5P.*LBI6);
  LBI8=(LBI7==1);
  [oo,pp]=find(LBI8==1);
  11=[mean(pp), mean(oo)];
  figure; imshow(LGRAY,[]); colormap('jet'); hold on
text(1213(1),1213(3),'+','Color',
[1 1 1],'FontSize',28,'HorizontalAlignment','center');
text(1213(2),1213(3),'+','Color',
[1 1 1],'FontSize',28,'HorizontalAlignment','center');
text(11(1),11(2),'+','Color',
[1 1 1],'FontSize',28,'HorizontalAlignment','center');
text(14(1),14(2),'+','Color',
[1 1 1],'FontSize',28,'HorizontalAlignment','center');
```

The results of detecting shoulder appendages (points l_1 and l_4) are marked with white points in Fig. 5.7.

In the fourth stage, the position of hips is detected—points l_6, l_7 (see Fig. 5.1). This detection starts with searching for the value of the row coordinate of the points l_6, l_7. For this purpose, a temperature gradient is analysed in the L_{ROIB} area being within the range of all the rows and columns $(n_{med} - p_{rb}, n_{med} + p_{rb})$—as shown in Fig. 5.8.

For the area of paraspinal muscles being in the range of ± 7 cm, the area visible in the image L_{GRAY}, whose resolution $M \times N = 140 \times 140$ pixels, is $p_{rb} = 10$ pixels. Thus, the source code fragment enabling to calculate the row coordinates of the points l_6 and l_7 16 is as follows:

```
prb=10;
[oo,pp]=find(sum(LBI3')>0);
LROIB=LGRAY(pp(length(pp)):end,(n_med-prb):(n_med+prb));
if max(abs(gradient(mean(LROIB'))))>0.9
    [oo,pp]=find(abs(gradient(mean(LROIB')))==max
(abs(gradient(mean(LROIB')))));
    1617(3)=round(size(LGRAY,1)-(length(mean(LROIB'))-pp(1)))-2;
else
    1617(3)=size(LGRAY,1);
end
```

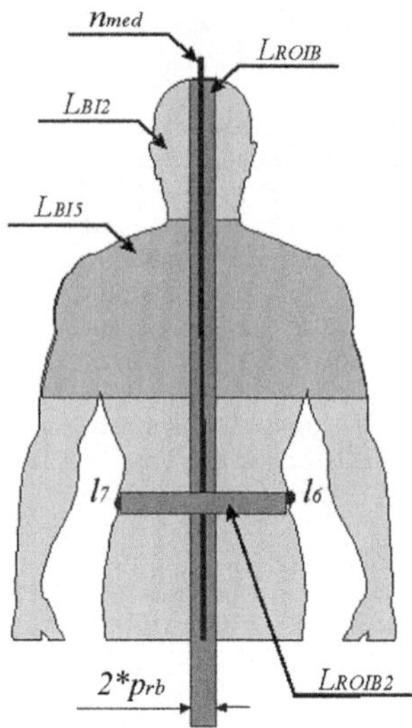

Fig. 5.8 Schematic diagram of the method for calculating the position of the points l_6 and l_7

The above method for calculating the gradient, on whose basis the position of individual points is determined, was also applied to the column axis.

$$L_{ROI2G}(m) = \frac{1}{N_{ROI2}} \left(\sum_{n=1}^{N_{ROI2}} L_{ROI2}(m,n) - \sum_{n=1}^{N_{ROI2}} L_{ROI2}(m+1,n) \right) \qquad (5.15)$$

where N_{ROI2} and M_{ROI2}—the number of columns and rows of the image L_{ROI2} respectively, and $m \in (1, M_{ROI2} - 1)$.

While detecting the position of the points l_6 and l_7 in the column axis, the search area was also adopted (defined by the parameter p_{rb}). The cut-off threshold for the waveform L_{ROI2G}, determined by the threshold p_{rgb}, was also added, i.e.:

```
prbg=0.6;
LROIB2=LGRAY((1617(3)-prb):(1617(3)+2),:);
LROI2G=abs(gradient(mean(LROIB2)));
figure; plot(LROI2G,'-g*'); hold on; grid on
xlabel('m [pixel]','FontSize',15,'FontAngle','Italic')
ylabel('L_{ROI2G}(m)','FontSize',15,'FontAngle','Italic')
[~,ppl]=find((LROI2G(1:n_med  )>prbg)==1);
[~,ppp]=find((LROI2G(n_med:end)>prbg)==1);
1617(1:2)=[ppl(length(ppl)),(ppp(2)+n_med)];
figure; imshow(LGRAY,[]); colormap('jet'); hold on
text(1213(1),1213(3),'+','Color',
[1 1 1],'FontSize',28,'HorizontalAlignment','center');
text(1213(2),1213(3),'+','Color',
[1 1 1],'FontSize',28,'HorizontalAlignment','center');
text(11(1),11(2),'+','Color',
[1 1 1],'FontSize',28,'HorizontalAlignment','center');
text(14(1),14(2),'+','Color',
[1 1 1],'FontSize',28,'HorizontalAlignment','center');
text(1617(1),1617(3),'+','Color',
[1 1 1],'FontSize',28,'HorizontalAlignment','center');
text(1617(2),1617(3),'+','Color',
[1 1 1],'FontSize',28,'HorizontalAlignment','center');
```

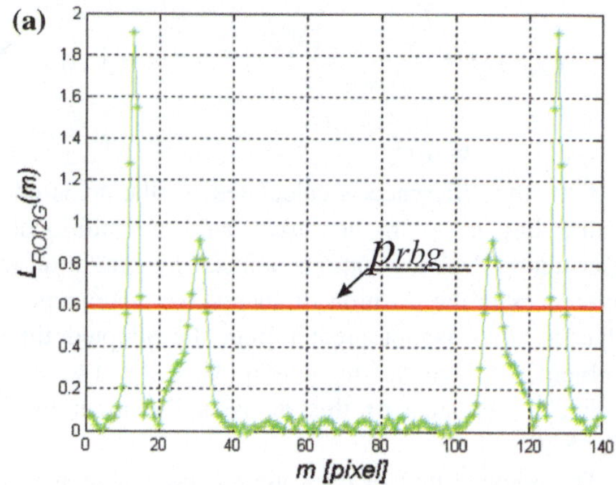

Fig. 5.9 Waveform of changes in the mean gradient in the area L_{ROIB2} **a** and **b** results of detection of the characteristic points

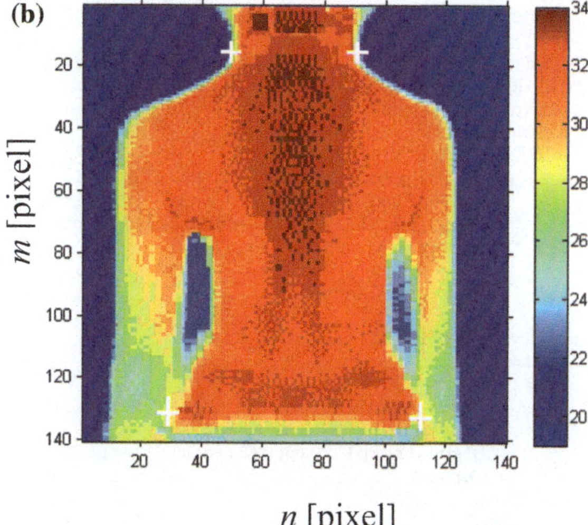

The results are shown in Fig. 5.9.

In the fifth stage, the position of the contours connecting the various points from l_1 to l_8 is detected. The points of the contour connecting l_5 and l_6 as well as l_7 and l_8 are detected using the gradient function searching for the next contour points starting with the position of l_6 and l_7 for the left and right side of the patient respectively.

For each row, starting from the bottom, the difference in temperature values for successive rows is calculated sequentially, i.e.:

$$L_{ROI3G}(m) = \frac{1}{N_{ROIB2}} \left| \sum_{n=1}^{N_{ROIB2}} L_{ROIB2}(m-1,n) - \sum_{n=1}^{N_{ROIB2}} L_{ROIB2}(m,n) \right| \qquad (5.16)$$

where $m,n \in ROIB2$.

Then the difference is calculated for the areas on the right and left side of the central line n_{med}. In the next stage, an additional condition is checked when determining the next contour points. This condition involves checking whether the differences in the position of successive contour points in the column axis for the adjacent rows is not greater than the adopted threshold p_{rt}. For the presented problem and the image resolution ($M \times N = 140 \times 140$ pixels), this is the value of 3 pixels. The idea of this analysis is shown in Fig. 5.10a and the results in Fig. 5.10b.

The relevant part of the source code is shown below:

```
prt=3;
LROI3G=abs(gradient(mean(  LGRAY(((1617(3)-2)-1):(1617(3)-2),:)  )));
[~,ppl]=find((LROI3G(1:n_med  )>prbg)==1);
[~,ppp]=find((LROI3G(n_med:end)>prbg)==1);
Lbokl=ppl(length(ppl));
Lbokp=ppp(2)+n_med;
Lbokl2=[];
for i=(1617(3)-2):-1:2
    LROI3G=abs(gradient(mean(  LGRAY((i-1):i,:)  )));
        [~,ppl]=find((LROI3G(1:n_med  )>prbg)==1);
        if (abs(Lbokl(1)-ppl(length(ppl))))>prt)
            break
        end
        Lbokl=[ppl(length(ppl)),i];       Lbokl2=[Lbokl2;Lbokl];
end
Lbokl2=flipud(Lbokl2);
Lbokp2=[];
for i=(1617(3)-2):-1:2
    LROI3G=abs(gradient(mean(  LGRAY((i-1):i,:)  )));
        [~,ppp]=find((LROI3G(n_med:end)>prbg)==1);
        if (abs(Lbokp(1)-(ppp(2)+n_med))>prt)
            break
        end
        Lbokp=[(ppp(2)+n_med),i]; Lbokp2=[Lbokp2;Lbokp];
end
Lbokp2=flipud(Lbokp2);
text(Lbokl2(:,1),Lbokl2(:,2),'+','Color',
[1 1 1],'FontSize',18,'HorizontalAlignment','center');
text(Lbokp2(:,1),Lbokp2(:,2),'+','Color',
[1 1 1],'FontSize',18,'HorizontalAlignment','center');
```

The above contour analysis applies only to the pairs of points l_5 and l_6 as well as l_7 and l_8. The other contour points are analysed in a similar manner (by calculating the derivative for each row and analyzing the shift values—of the adopted threshold p_{rt}). The results obtained are shown in Fig. 5.8a.

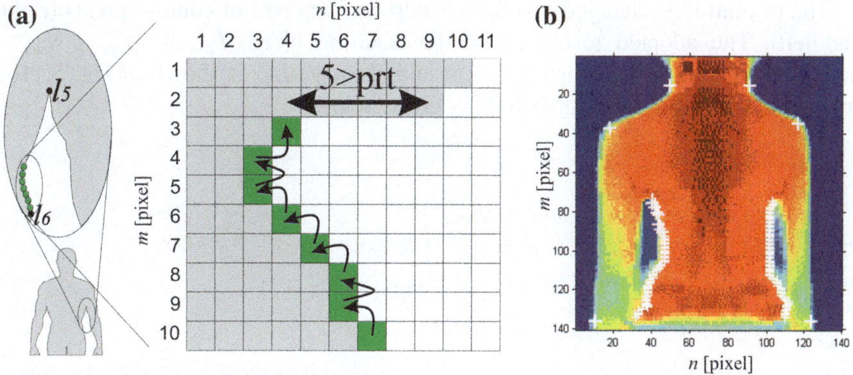

Fig. 5.10 Schematic diagram and the results of connecting individual points: **a** explanation of the method for connecting points l_5 and l_6; **b** examples of the results marked in *white* in the infrared image

The source code in this case is as follows:

```
LROI4G=abs(gradient(LGRAY(l213(3),:)));
Lbokl3=[];
for  i=l213(3)+2:11(2)
    LROI4G=abs(gradient(LGRAY(i,:)   ));
    [jednynki,ppl]=find((LROI4G(1:n_med   )>0.5)==1);
    Lbokl=[ppl(length(ppl)),i];
    Lbokl3=[Lbokl3;Lbokl];
end
Lbokp3=[];
for  i=l213(3)+2:14(2)
    LROI4G=abs(gradient(LGRAY(i,:)   ));
    [jednynki,ppp]=find((LROI4G(n_med:140)>0.5)==1);
    Lbokp=[(ppp(2)+n_med),i];
    Lbokp3=[Lbokp3;Lbokp];
end
figure; imshow(LGRAY,[]); colormap('jet'); hold on
text(Lbokl3(:,1),Lbokl3(:,2),'+','Color',
[1 1 1],'FontSize',18,'HorizontalAlignment','center');
text(Lbokp3(:,1),Lbokp3(:,2),'+','Color',
[1 1 1],'FontSize',18,'HorizontalAlignment','center');
Lbokl4_n=Lbokl3(length(Lbokl3),2):Lbokl2(1,2);
Lbokl4_m=round(interp1([Lbokl3(length(Lbokl3),2),
Lbokl2(1,2)],[Lbokl3(length(Lbokl3),1),Lbokl2(1,1)],
Lbokl4_n ));
Lboklall=[Lbokl3;[Lbokl4_m',Lbokl4_n'];Lbokl2];
Lbokp4_n=Lbokp3(length(Lbokp3),2):Lbokp2(1,2);
Lbokp4_m=round(interp1([Lbokp3(length(Lbokp3),2),
Lbokp2(1,2)],[Lbokp3(length(Lbokp3),1),
Lbokp2(1,1)],Lbokp4_n ));
Lbokpall=[Lbokp3;[Lbokp4_m',Lbokp4_n'];Lbokp2];
figure; imshow(LGRAY,[]); colormap('jet'); hold on
text(Lboklall(:,1),Lboklall(:,2),'0','Color',
[0 0 0],'FontSize',18,'HorizontalAlignment','center');
text(Lbokpall(:,1),Lbokpall(:,2),'0','Color',
[0 0 0],'FontSize',18,'HorizontalAlignment','center');
```

The resolution is changed in a further step for each pair of contour points (right and left). The adopted resolution of the resulting image L_{net} is $M_{net} \times N_{net} = 111 \times 100$ pixels. An example of the resulting image L_{net} is shown in Fig. 5.11b, while the corresponding source code is given below.

```
Lnet=[]; Nnet=100;
for ii=1:size(Lbokpall,1)
    Lnet(ii,1:Nnet)=imresize(LGRAY
(Lboklall(ii,2),Lboklall(ii,1):Lbokpall(ii,1)),[1 Nnet]);
end
figure; imshow(Lnet,[min(LGRAY(:)) max(LGRAY(:))]);
colormap('jet'); colorbar
```

The image L_{net} thus created allows for any statistical calculations and analyses. The aforementioned approach is the analysis of the mean temperature in the areas designated by raster shown in Fig. 5.1. Calculation of the mean, minimum or maximum values is now possible for any raster size and independent of the patient size. This is a crucial advantage during the analysis of the spine and temperature deviations from the norm. The results of analysis in the raster areas (matrix L_{mac}) are shown in Fig. 5.2. The source code providing the matrix L_{mac} and Fig. 5.2 is shown below:

```
figure('Color',[1 1 1]);
imshow(LGRAY,[],'notruesize'); colormap('jet'); colorbar;
hold on
Lmac=[];
krok_y=(length(Lboklall)/21);
for ij=1:krok_y:(length(Lboklall)-krok_y)
    ij=round(ij);
line([Lboklall(ij,1),Lbokpall(ij,1)],[Lboklall(ij,2),
Lbokpall(ij,2)],'Color',[1 1 1],'LineWidth',2);
    ggp=Lboklall(ij,1):((Lbokpall(ij,1)-
Lboklall(ij,1))/8):Lbokpall(ij,1);
    ggw=Lboklall(round(ij+krok_y),1):((Lbokpall(round(ij+krok_y),1)-
Lboklall(round(ij+krok_y),1))/8):Lbokpall(round(ij+krok_y),1);
    for g=1:(length(ggp)-1)
        line([ggp(g),
ggw(g)],[Lboklall(ij,2),Lboklall(round(ij+krok_y),2)],
'Color',[1 1 1],'LineWidth',2);
Lwyc=roipoly(LGRAY,[Lboklall(ij,2),Lboklall(ij,2),
Lboklall(round(ij+krok_y),2),Lboklall(round(ij+krok_y),2)],
[ggp(g),ggp(g+1),ggw(g+1),ggw(g)]);
        Lw=double(Lwyc').*LGRAY; Lw=Lw(:); Lw(Lw==0)=[];
        if ~isempty(Lw)
            Lmac(ij,g)=mean(Lw);
        else
            break
        end
    end
    line([ggp(length(ggp)),ggw(length(ggp))],[Lboklall(ij,2),
Lboklall(round(ij+krok_y),2)],'Color',[1 1 1],'LineWidth',2);
end
line([Lboklall(round(ij+krok_y),1),Lbokpall(round(ij+krok_y),1)],
[Lboklall(round(ij+krok_y),2),Lbokpall(round(ij+krok_y),2)],
'Color',[1 1 1],'LineWidth',2);
Lmac((Lmac(:,1)==0),:)=[];
```

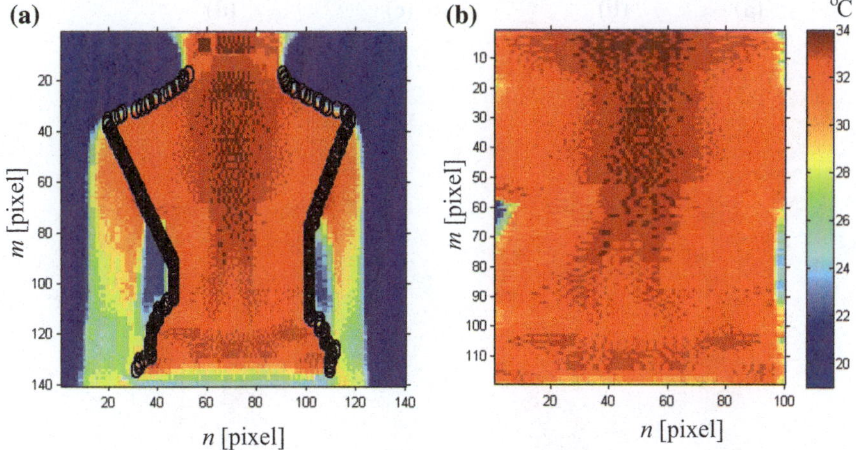

Fig. 5.11 Schematic diagram and the results of connecting individual points: **a** explanation of the method for connecting points l_5 and l_6; **b** examples of the results marked in *white* in the infrared image

The resulting matrix L_{mac} is another objective that has been achieved. This matrix can be further analysed and, above all, because of its small size, archived. The matrix L_{net} (containing much more information compared to the matrix L_{mac}) enables to estimate the position of the spine. Until now, the measurement of spinal curvature on X-ray images has been performed using three known methods by Gruca, Fergusson and Cobb—according to Fig. 5.12.

Fig. 5.12 Methods for measuring spinal curvatures:
a Gruca's method;
b Fergusson's method;
c Cobb's method

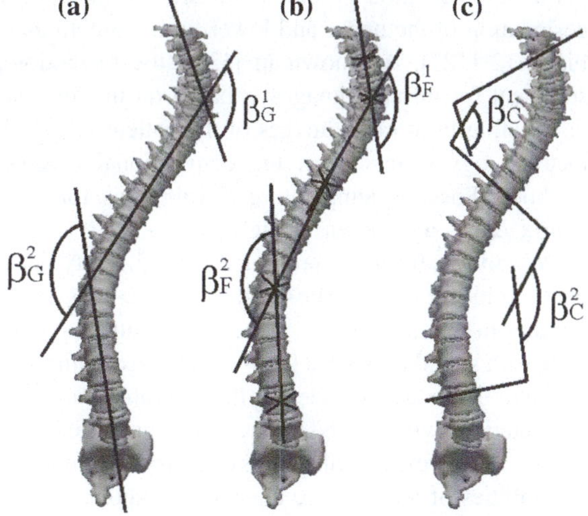

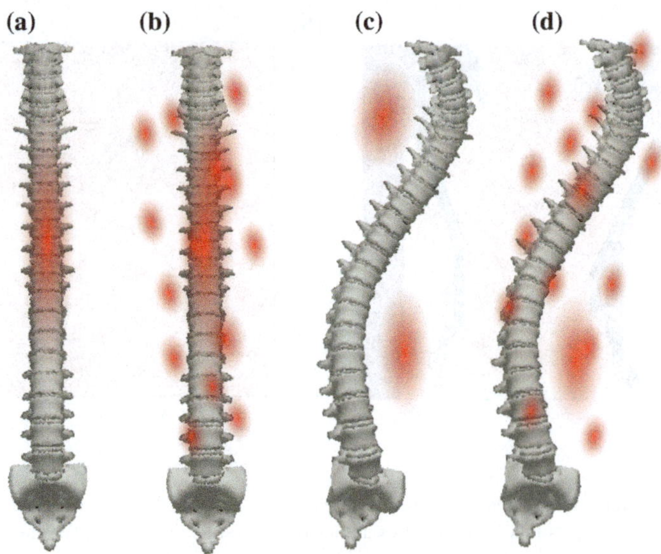

Fig. 5.13 Classical temperature distributions for patients: **a** healthy subject BMI $\in$ (18.5, 25); **b** healthy subject BMI > 25; **c** ill patient BMI $\in$ (18.5, 25); **d** ill patient BMI > 25

The presented methodology for measuring spinal curvature generally involves plotting straight lines along the top surface of the upper extreme vertebra of the curvature and the lower surface of the lower extreme vertebra. In the case of Cobb's method, this is the measurement of the angle of intersection of the lines perpendicular to the above straight lines. In the case of Gruca's method, lines are drawn from the root of curvature arcs of the concave side to the root of three top vertebrae. Fergusson's method is similar to Gruca's method. The difference is only in the designation of the upper and lower limit vertebra as well as the atlas vertebra—see Fig. 5.12 [127]. As shown in [133], the methodology of spinal curvature measurement for infrared images is based on the analysis of clusters and temperature distribution in infrared images of the patient's back. In this case, the measurement methodology differs from the conventional (shown in Fig. 5.12) measurement methods. Possible temperature distributions for the matrix L_{net} and the original matrix L_{GRAY} are shown in Fig. 5.13.

As follows from the presented (Fig. 5.13) typical temperature distributions for obese subjects (Body Mass Index, BMI > 25), it is difficult to predict the spine outline. In practice, these subjects constitute a group of exclusion. For other subjects with spinal curvature (above 10 [°] according to Cobb), a temperature increase is from the concave side of the curvature arc—as shown in Fig. 5.13c. This characteristic will be used to determine the outline of the spine in the infrared image L_{net}. A key element is also the symmetry of temperature distribution. Three typical possibilities of temperature changes are shown in Fig. 5.14.

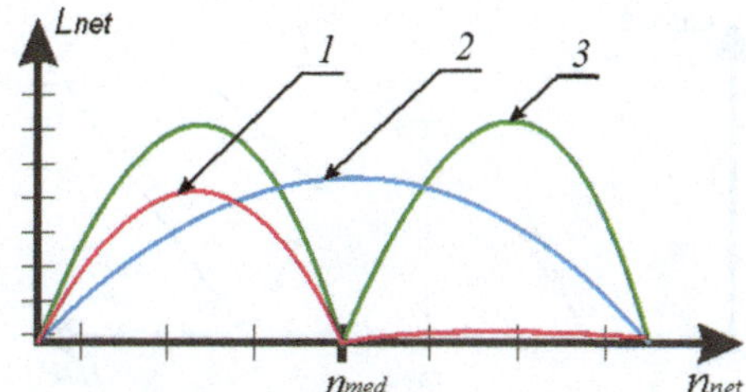

Fig. 5.14 Illustrative graph of temperature changes of L_{net} for the next columns n. The colours indicate: *1* spinal curvature—unilateral temperature change; *2* maximum temperature centrally in the middle—healthy subject; *3* symmetry in temperature changes—healthy subject

Figure 5.14 shows a graph of temperature changes of L_{net} for spinal curvature, Fig. 5.14(1) (unilateral temperature change), and temperature changes for healthy subjects, Fig. 5.14(2, 3) (symmetric temperature change). Therefore a method was proposed for calculating the gradient of temperature changes for each of the rows from the designated centre n_{med}. This method involves calculating the value of the gradient of $L_{netG2}(m_{net}, n_{net})$ defined as follows:

$$L_{netG2}(m_{net}, n_{net}) = \begin{cases} L_{netG}(m_{net}, n_{net}) & if & (L_{netG}(m_{net}, n_{net}) > 0) \wedge (n < N_{net}/2) \\ -L_{netG}(m_{net}, n_{net}) & if & (L_{netG}(m_{net}, n_{net}) < 0) \wedge (n \geq N_{net}/2) \\ 0 & other \end{cases}$$

(5.17)

$$L_{netG}(m_{net}, n_{net}) = L_{net}(m_{net}, n_{net}) - L_{net}(m_{net}, n_{net} + 1)$$ (5.18)

for: $n_{net} \in (1, N_{net} - 1)$.

The values of the gradient of $L_{netG2}(m_{net}, n_{net})$ change their sign after crossing the position of the spine in successive rows. For the other columns, the values of $L_{netG2}(m_{net}, n_{net})$ are equal to zero—Fig. 5.15 (compare with Fig. 5.14). The corresponding source code allowing for these transformations is as follows:

```
LnetG=diff(Lnet,1,2);
figure; imshow(LnetG,[]); colormap('jet')
Lnet1G=LnetG(:,1:round(size(LnetG,2)/2));
LnetpG=LnetG(:,round(size(LnetG,2)/2):end);
Lnet1G(Lnet1G>0)=0;
LnetpG(LnetpG<0)=0;
LnetG2=abs([Lnet1G,LnetpG]);
figure; imshow(LnetG2,[]); colormap('jet')
```

In the next step, the matrix $L_{netG2}(m_{net}, n_{net})$ is subjected to filtration with an averaging filter. In this case, the mask size of the filter h_{unet} may be changed in a narrow range due to the possibility of blurring the edge of individual temperature changes. This is the range for the analysed resolution of the image L_{GRAY}, namely

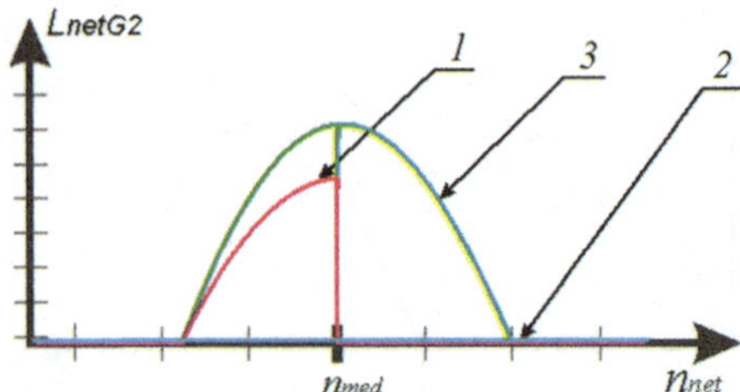

Fig. 5.15 Illustrative graph of changes in the directional derivative of temperature of $L_{netG2}(m_{net}, n_{net})$ for the next columns n. The colours indicate: *1* spinal curvature—unilateral temperature change; *2* maximum temperature centrally in the middle—healthy subject; *3* symmetry in temperature changes—healthy subject

$M_{unet} = N_{unet} \in (3,21)$ pixels. The matrix $L_{netG4}(m_{net}, n_{net})$ after filtration with an averaging filter is therefore equal to:

```
LnetG3=conv2(LnetG2,ones(19),'same');
LnetG4=LnetG3./sum(sum(ones(19)));
```

Next, the matrix is normalized. The output matrix $L_{netG4n}(m_{net}, n_{net})$ is therefore equal to:

$$L_{netG4n}(m_{net}, n_{net}) = \frac{\left(L_{netG4n}(m_{net}, n_{net}) - \min_{m_{net}, n_{net}} L_{netG4n}(m_{net}, n_{net}) \right)}{\max_{m_{net}, n_{net}} \left(L_{netG4n}(m_{net}, n_{net}) - \min_{m_{net}, n_{net}} L_{netG4n}(m_{net}, n_{net}) \right)} \quad (5.19)$$

The corresponding source code based on the function `mat2gray` (a simplified version of the function `imadjust`) is as follows:

```
LnetG4n=mat2gray(LnetG4);Lgauss3=[];
```

The interpretation shown in Fig. 5.16 was proposed for the analysis of the temperature distribution in the paraspinal area. The curvature of the spine is shown here with heat sources as a deviation from the correct position. It can be assumed here that each of the heat sources has a centre (e.g. maximum temperature). This centre is connected by means of a hypothetical spring with the vertebra (atlas vertebrae). Any increase in temperature (especially in the paraspinal area) is another spring pushing away the spine (see Fig. 5.16).

It is a difficult task to write the hypothesis presented in Fig. 5.16 in the form of an algorithm or a mathematical formula. This is due to the need to take into account the position of the heat source, the degree of influence on the outline of the spine

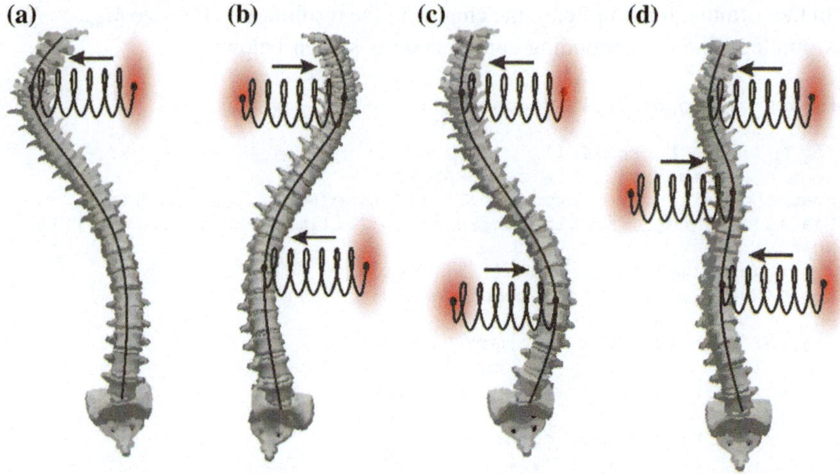

Fig. 5.16 Interpretation of temperature changes and the position of a hypothetical spring causing deformation of the spine

and others. This task starts by defining two Gaussian distributions, i.e. matrices L_{gauss} and L_{gauss2}:

$$L_{gauss}(m,n) = \frac{exp\left(\frac{-\left(n\cdot\frac{1}{4}\cdot\frac{1}{M\cdot N}\left(L_{bokpall}(m,1)-L_{boklall}(m,1)\right)\right)^2}{2\cdot p_{std}^2}\right)}{p_{std}\cdot\sqrt{2\cdot\pi}} \tag{5.20}$$

$$L_{gauss2}(m_{net},n_{net}) = \frac{exp\left(\frac{-\left(\left(n_{net}-\frac{N_{net}}{2}\right)\cdot\frac{2}{N_{net}}\right)^2}{2\cdot p_{std2}(m_{net})^2}\right)}{p_{std2}(m_{net})\cdot\sqrt{2\cdot\pi}} \tag{5.21}$$

where p_{std}—standard deviation of the mean equal to 0.3,
   p_{std2}—standard deviation of the mean equal to:

$$p_{std2}(m_{net}) = \frac{1}{N_{net}}\sum_{n_{net}=1}^{N_{net}} L_{netG4n}(m_{net},n_{net}) \tag{5.22}$$

for $m \in (1,M)$.

In the formula, for simplicity, the change in the resolution to the size $M_{net} \times N_{net}$ was omitted. The corresponding source code is shown below:

```
Lgauss=[]; Lgauss2=[]; Lgauss3=[];
pstd=0.3;
for m=1:size(Lbokpall,1)
    Lngauss=gauss( ((1:size(LGRAY,2))-
mean(Lboklall(m,1):Lbokpall(m,1)))./(size(LGRAY,2)/2),pstd);
Lgauss(m,1:Nnet)=imresize(Lngauss(Lboklall(m,1):Lbokpall(m,1)),[1
Nnet] );
    Lgauss2(m,1:Nnet)=gauss( ((1:Nnet)-
round(Nnet/2))./(Nnet/2),mean(LnetG4n(m,:))));
end
figure; mesh(Lgauss); colormap('jet');
xlabel('n_{net} [pixel]','FontSize',18,'FontAngle','Italic');
ylabel('m_{net} [pixel]','FontSize',18,'FontAngle','Italic');
zlabel('L_{gauss} [pixel]','FontSize',18,'FontAngle','Italic');
figure; mesh(Lgauss2); colormap('jet');
xlabel('n_{net} [pixel]','FontSize',18,'FontAngle','Italic');
ylabel('m_{net} [pixel]','FontSize',18,'FontAngle','Italic');
zlabel('L_{gauss2} [pixel]','FontSize',18,'FontAngle','Italic');
figure; mesh(Lgauss+Lgauss2/4); colormap('jet');
xlabel('n_{net} [pixel]','FontSize',18,'FontAngle','Italic');
ylabel('m_{net} [pixel]','FontSize',18,'FontAngle','Italic');
zlabel('L_{gauss}.*L_{gauss2}[pixel]','FontSize',18,
'FontAngle','Italic');
```

where gauss is the following function:

```
function y = gauss(x,std)
y = exp(-x.^2/(2*std^2)) / (std*sqrt(2*pi));
```

The resulting matrices L_{gauss} and L_{gauss2} enable to take into account the weight of different temperature foci. The matrix L_{gauss}, according to the Gaussian distribution, changes the weights of temperature resulting from the need to expand (change the resolution) to the fixed size of the matrix L_{net}. Therefore temperature changes close to the position of the spine are preferred (according to the Gaussian distribution). The further away from the central position, the smaller effect a given temperature focus has on the outline of the spine. The matrix L_{gauss2} in turn affects the weight taking into account the temperature values. High temperature values affect the result to a greater extent than low values. This implies the use of both matrices for subsequent analysis, i.e. L_{netG5} (analogously to the Eq. 5.19):

$$L_{netG5}(m_{net}, n_{net}) = \frac{\left(L_{netG5n}(m_{net}, n_{net}) - \min_{m_{net}, n_{net}} L_{netG5n}(m_{net}, n_{net}) \right)}{\max_{m_{net}, n_{net}} \left(L_{netG5n}(m_{net}, n_{net}) - \min_{m_{net}, n_{net}} L_{netG5n}(m_{net}, n_{net}) \right)} \quad (5.23)$$

$$L_{netG5n}(m_{net}, n_{net}) = L_{netG4}(m_{net}, n_{net}) \cdot \left(L_{gauss}(m_{net}, n_{net}) + \frac{1}{4} \cdot L_{gauss2}(m_{net}, n_{net}) \right)$$

$$(5.24)$$

The above formula was written in Matlab as:

```
LnetG5=mat2gray(LnetG4.*(Lgauss+Lgauss2/4));
```

The relations (5.23) and (5.24) as well as the weight (the number "4") were adopted on the basis of experience and analyses performed on the collected examples. Both the above relations and the weight can be freely modified by the reader (to which I encourage). However, the impact on the obtained results should be analysed quantitatively for each case and relation. The matrices L_{gauss} and L_{gauss2} are shown on the analysed example in Fig. 5.17.

After taking into account the weight contribution of individual temperature foci in the outline of the spine, its position can be analysed. To do this, each of the rows of the matrix L_{netG5} must be analysed separately. This analysis involves designation of such a position of a single point for each row, for which the difference in mean temperature values is the smallest. The mean temperature values will be calculated from the left and right sides of the analysed point of the matrix L_{netG5}. The idea of the calculation is shown in Fig. 5.18.

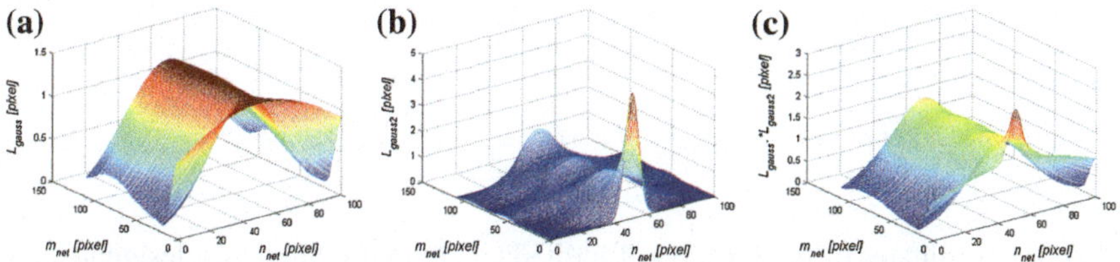

Fig. 5.17 Matrices: **a** $L_{gauss}(m_{net}, n_{net})$; **b** $L_{gauss2}(m_{net}, n_{net})$; and **c** $L_{gauss}(m_{net}, n_{net}) \cdot L_{gauss2}(m_{net}, n_{net})$

Fig. 5.18 Explanatory figure of searching for a minimum difference in mean temperature values for the next rows of the matrix L_{netG5} (after scaling to the matrix L_{GRAY}). The *red line* indicates the detected fragment of the spine

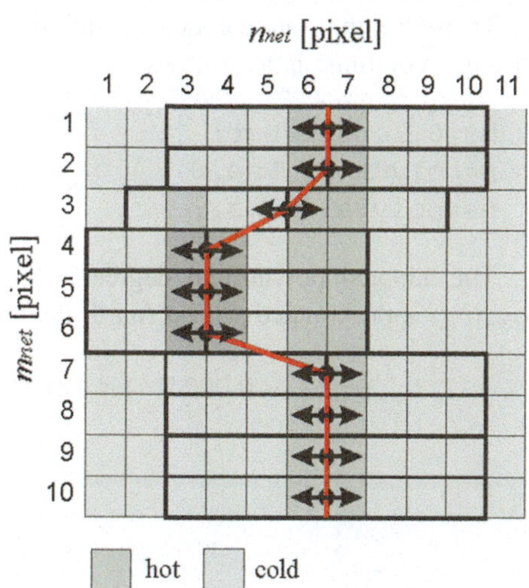

For each row of the matrix L_{netG5} the absolute difference of mean temperature values is stored in the sub-matrix *pam_s*, i.e.:

$$pam\_s(n_{net}) = \frac{1}{N_{net}} \left| \sum_{n_{net}=\frac{N_{net}}{2}}^{\left|\frac{N_{net}}{2}+\frac{N_{net}}{8}\right.} L_{netG5}(m_{net}, n_{net}+n_{var}) \right. $$
$$\left. - \sum_{n_{net}=\frac{N_{net}}{2}-\frac{N_{net}}{8}}^{\frac{N_{net}}{2}} L_{netG5}(m_{net}, n_{net}+n_{var}) \right| \tag{5.25}$$

for $n_{var} \in (-N_{net}/8, N_{net}/8)$.

This means that each of the contour points of the spine can be moved from right to left in the range of $\pm N_{net}/8$ (see Fig. 5.18, black circles and red outline of the spine). The relevant fragment is as follows:

```
Nt=round(size(LnetG5,2)/2);
LnetG6=zeros(size(LnetG5));
Lsp=[]; Nz=round(size(LnetG5,2)/8);
for m=1:size(LnetG5,1)
    pam_s=[];
    for n=(Nt-round(size(LnetG5,2)/8)):(Nt+round(size(LnetG5,2)/8))
        pam_s=[pam_s;[n,abs(mean(LnetG5(m,(n-Nz):(n)))-
mean(LnetG5(m,(n):(n+Nz))))]];
    end
    pam_s_s=sortrows(pam_s,2);
    Lsp(m)=pam_s_s(1,1);
end
figure; imshow(LnetG4.*Lgauss,[]); colormap('jet'); hold on
plot(Lsp,1:length(Lsp),'*k')
```

Thus coordinates of the position of individual contour points thus designated are stored in the variable L_{sp}. These points require filtration and/or polynomial approximation. This will allow for automatic determination of inflection points and tangents at these points according to Gruca's measuring method (see Fig. 5.12).

Therefore, the source code of eighth degree polynomial interpolation (m_s—rows and n_s—columns) is as follows:

```
ms=0:length(Lsp)-1;
p=polyfit(ms,Lsp,8);
ns=polyval(p,ms);
```

The adopted polynomial degree is 8. For example, the outline of the spine $L_{spw}(m_s)$ approximated with a fourth degree polynomial is as follows:

$$L_{spw}(m_s) = -0.00241 \cdot m_s^3 + 0.04 \cdot m_s^2 - 0.23 \cdot m_s + 49.21 \tag{5.26}$$

The adopted polynomial degree (8) is the result of the author's previous experience and comparisons of possible spinal curvatures with polynomial functions. It means that the reader can test the effect of different degrees of the polynomial on the results obtained. On this basis, the next, and also the last step is calculating the inflection points and designation of tangents at these points. The inflection points are determined using the function `roots`, i.e.:

```
p3=[]; p4=[];
ms2=1:length(p);
p2=(length(p)-ms2(:)).*p(:);
p3(1:length(p)-1)=p2(1:length(p)-1);
ms2=1:length(p3);
p5=(length(p3)-ms2(:)).*p3(:);
p4(1:length(p3)-1)=p5(1:length(p3)-1);
proots=roots(p4);
ri=1;rre=1;
mns=[ms(:),ns(:)];
figure('Color','w');
plot((mns(:,2)),mns(:,1),'-r*')
hold on
xlabel('n_{s} [pixel]','FontSize',18,'FontAngle','Italic');
ylabel('m_{s} [pixel]','FontSize',18,'FontAngle','Italic');
grid on
m_tang=[]; m_tang2=[]; dfg=[]; beta=[];
while(ri<=length(proots))
    if imag(proots(ri))==0
        if((proots(ri)<=length(Lsp)))
           dfg(rre,1:2)=mns(ceil(proots(ri)),:);
           plot(((dfg(rre,2))),dfg(rre,1),'o');
           m_tang2(rre)=polyval(p3,dfg(rre,1));
           tti=(dfg(rre)-20):(dfg(rre)+20);
           m_tang(rre,1:length(tti))=m_tang2(rre)*
(tti-dfg(rre,1))+dfg(rre,2);
           plot((m_tang(rre,:)),tti,'.g');grid on
           if(rre>1)
beta(rre)=abs(atan(m_tang2(rre))*180/pi)+
abs(atan(m_tang2(rre-1))*180/pi);
             text((dfg(rre,2)),(dfg((rre-1),1)-
dfg((rre),1))/2+dfg(rre,1),[num2str(round((beta(rre))*
10)/10),'^o']);
           end
           rre=rre+1;
        end
    end
    ri=ri+1;
end
beta
```

The resulting graph is shown in Fig. 5.19. The designated tangents (marked in green in Fig. 5.19) enable in this case to calculate automatically three angular values $\beta_G^1 = 4.69$ [°], $\beta_G^2 = 4.88$ [°] and $\beta_G^3 = 9.18$ [°]. For this purpose, the roots were analysed—values of the variable `proots`

```
86.797
49.165 +    36.304i
49.165 -    36.304i
 66.23
36.318
10.615
```

Therefore, four real roots were found. For them, the tangents marked in green in Fig. 5.19 were calculated. The length of the tangents was determined (only for the purposes of visualization) as ±20 pixels. The combination of the input image L_{GRAY} and the calculated approximation of the spine outline is shown in the form of three-dimensional graphs in Fig. 5.20.

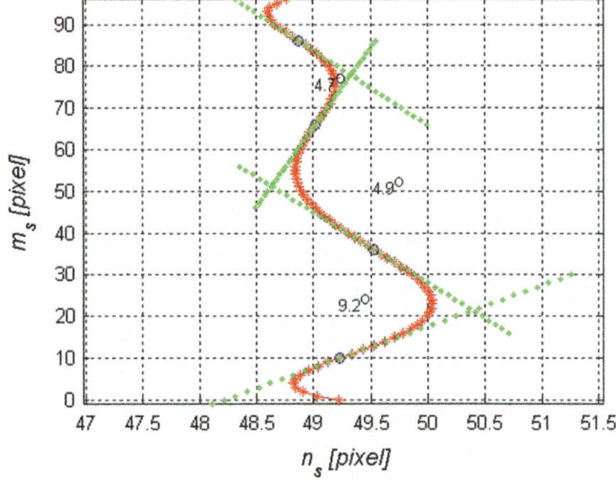

Fig. 5.19 Sample graph $L_{spw}(m_s)$ of the detected position of the spine—*red*, and tangents at the points of inflection—*green*. The angle of curvature was indicated for each pair of tangents automatically

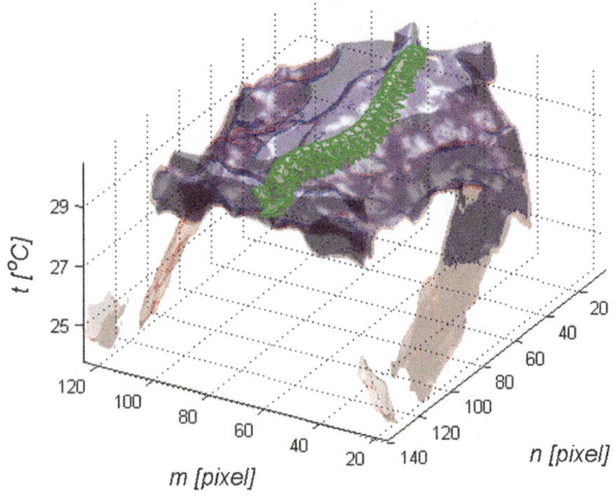

Fig. 5.20 Graph of areas of the matrix L_{GRAY} grouped according to the temperature values with the spine outline marked in *green*

The relevant fragment providing the results shown in Fig. 5.20 is given below:

```
jj=1; Lf=[];
for kk=23:0.1:max(LGRAY(:))
    Lf(:,:,jj)=((LGRAY>(kk-2))&(LGRAY<kk)).*LGRAY;
    jj=jj+1;
end
Lf = smooth3(Lf,'box',5);
figure
h=patch(isosurface(Lf,34)); axis equal, view(3)
isonormals(Lf,h)
camlight, lighting gouraud,
alpha(h,0.4)
set(h,'FaceColor','green','EdgeColor','none');
h=patch(isosurface(Lf,30)); axis equal, view(3)
isonormals(Lf,h)
camlight, lighting gouraud,
alpha(h,0.3)
set(h,'FaceColor','blue','EdgeColor','none');

h=patch(isosurface(Lf,27)); axis equal, view(3)
isonormals(Lf,h)
camlight, lighting gouraud,
alpha(h,0.2)
set(h,'FaceColor','magenta','EdgeColor','none');

set(h,'FaceColor','red','EdgeColor','none');
xlabel('m [pixel]','FontSize',20,'FontAngle','Italic');
ylabel('n [pixel]','FontSize',20,'FontAngle','Italic');
zlabel('t [^oC]','FontSize',20,'FontAngle','Italic');grid on;hold on

view(-150,26)
set(gca,'ZTickLabel',{'25';'27';'29';'31';'33'})
```

The data obtained in this subchapter can be freely processed further. Particularly noteworthy is the image L_{net} which, for example, can be further analysed using the gray-level co-occurrence matrix (GLCM). The proposed source code is given below and the sample results are presented in Fig. 5.21.

```
glcm = graycomatrix(Lnet,'GrayLimits',[min(Lnet(:))
max(Lnet(:))],'Offset',[0 3],'NumLevels',40);
figure; imshow(glcm,[],'InitialMagnification','fit');
colormap('jet'); colorbar
xlabel('Temperature [^oC]','FontSize',20,'FontAngle','Italic');
ylabel('Temperature [^oC]','FontSize',20,'FontAngle','Italic');
axis([25 40.5 25 40.5])
```

The presented fragment is also the last stage of automatic determination of the angle of spinal curvature in infrared images. A simplified block diagram of the whole algorithm is shown in Fig. 5.22.

The simplified block diagram shown in Fig. 5.22 and the algorithm created on its basis are attached to the main menu of the application in the form of a button. Arrangement of the individual buttons tailoring the algorithm is presented at the end of this chapter.

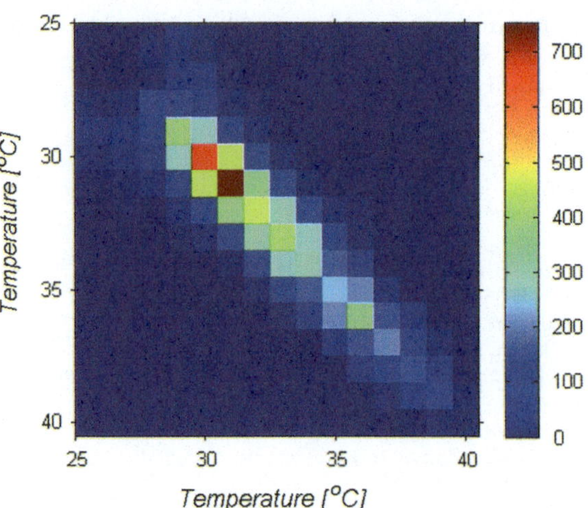

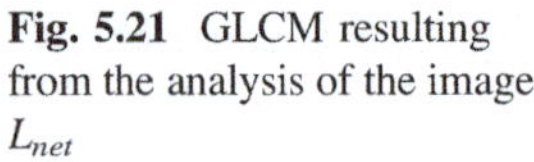

Fig. 5.21 GLCM resulting from the analysis of the image L_{net}

5.2   Quantitative Assessment of the Quality of Dermatological Laser Treatments

Dermatological treatments are largely associated with lasers as they are increasingly popular. In particular, ablative [134–136] and non-ablative lasers [137] are used to remove all types of skin discoloration or to make the skin smoother and firmer. The facial skin is most frequently subjected to laser treatments. In a typical procedure, the laser head is applied to the skin area undergoing therapy—e.g. the above-mentioned facial skin. Each laser application requires from the operator pressing the button triggering radiation. Typically, each radiation dose is adjacent to the previous one, i.e. the area of the skin subjected to treatment is in the shape of a square whose side is several millimetres long (depending on the type of laser used and the type of the head). For CO_2RE laser (CO_2 type—10,600 [nm] wave-length, 4.5 [J/cm^2] energy density, 1–150 [mJ] impulse energy, 16.7 [kHz] impulse frequency, 20–3000 [μs] impulse duration, 120 [μm] or 150 [μm] dot size), this is an area regulated by software and includes a square whose side is sized from 2 to 10 [mm] [138–141]. An example of performing the tests and measurements is shown in Fig. 5.23.

Laser treatments performed on the patient Fig. 5.23(1) by the operator Fig. 5.23 (2) are monitored with the infrared camera Fig. 5.23(9) used by the operator Fig. 5.23(8). Independently recorded data and laser settings are saved on the computer and processed using the algorithm presented in the following paragraphs.

The use of infrared cameras for this type of treatment includes two main areas:

- verification of the correctness of performing laser treatments—adjacency of individual radiation doses and
- determination of the type and shape of the laser impulse to optimize the treatment in terms of heating the patient's skin and the speed of the treatment.

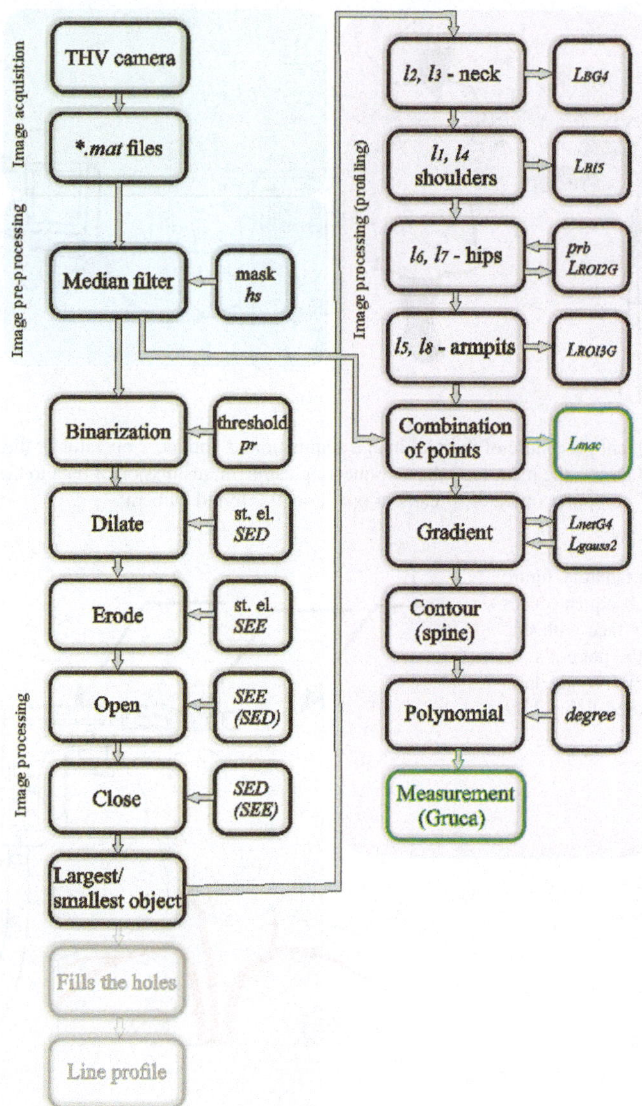

Fig. 5.22 Simplified block diagram of the algorithm for automatic analysis of the temperature distribution on the human back. The colours indicate four stages: image acquisition—*green*, image pre-processing—*blue*, image processing—*red* and image processing tailored to the calculations discussed in this subchapter—*red*. The text in *green* indicates significant results

Both the first as well as the second research area need to be monitored during and after treatment of the skin area. This monitoring requires tailoring of the application presented in the previous chapter. A detailed description of the tailored application, the algorithm for image analysis and processing, is presented in the following paragraphs.

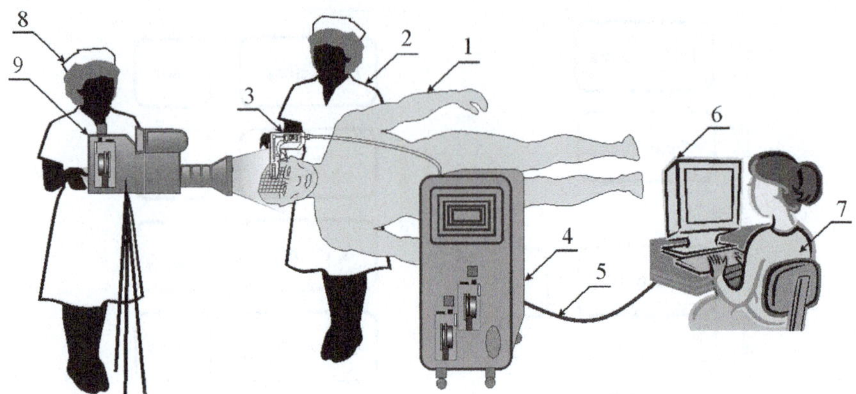

Fig. 5.23 Typical procedure of thermal image acquisition: *1* patient; *2* operator of the laser head; *3* laser head; *4* laser—the main module; *5* connecting cable—transmission of data to the computer; *6* computer; *7* computer operator; *8* camera operator; *9* infrared camera

Fig. 5.24 Explanatory figure of applying subsequent doses to the patient's face with the laser head: *1* the patient's head; *2* marks left by the laser visible in infrared light; *3* the laser head

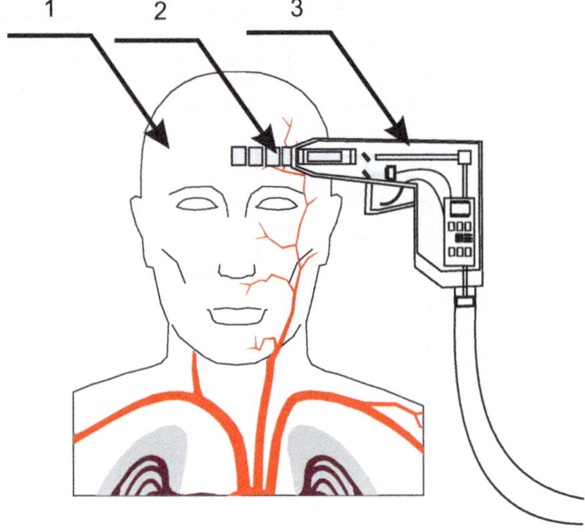

The main task of the application is to detect the places where the laser has already been applied based on temperature changes Fig. 5.24. As mentioned earlier, due to the square shape of the area, the whole procedure boils down to finding blurred squares on the skin. During the search, it is assumed that the infrared camera is set at a right angle relative to the skin area. However, this assumption cannot always be met, e.g. during the most common facial treatments. In the case of the nose, eyes and cheeks, the camera arranged perpendicularly to the patient's face shows these areas at a different angle. However, in practice, it is not a significant contraindication and there is no need to refer to more sophisticated methods of reconstruction and perspective correction.

Figure 5.25 shows a sample image L_{GRAY} with automatically marked centres of individual laser effects. It is possible thanks to the following fragment:

```
LGRAY=zeros(100);
R=round(rand([8 2])*(size(LGRAY,1)-1))+1;
for r=1:size(R,1)
    LGRAY(R(r,1),R(r,2))=1;
end
LGRAY2 = bwdist(LGRAY,'cityblock');
figure; imshow(LGRAY2,[])
prg=14;
figure; imshow(LGRAY2<prg)
h=ones(9);
LGRAY3=conv2(double(LGRAY2<prg),h,'same');
figure; imshow(LGRAY3,[])
LGRAY4=imopen(LGRAY3,ones(9));
LGRAY5=imregionalmax(LGRAY4);
LLABEL=bwlabel(LGRAY5);
figure; imshow(LGRAY4,[]); hold on
for i=1:max(LLABEL(:))
    Li=LLABEL==i;
    [xx,yy]=sr_ciez(Li);
    plot(xx,yy,'r*')
end
```

where `sr_ciez` is the function calculating the centre of gravity of the object (or objects) visible in the input image, i.e.:

```
function [xx,yy]=sr_ciez(L)
[m,n]=size(L);
La=ones(m,1)*(1:n);
L3a=La.*L;
Lb=(1:m)'*ones(1,n);
L3b=Lb.*L;
xx=sum(L3a(:))/sum(sum(L));
yy=sum(L3b(:))/sum(sum(L));
```

As seen in Fig. 5.25, detection of the areas where the laser was applied is easy when they are isolated. Since the main purpose of the method is also the detection of overlapping areas, a modification needs to be made. It involves taking into account the mean temperature in the area where the laser was applied. An example of such an area immediately after laser treatment is shown in Fig. 5.26.

For further analysis, it was assumed that the input image is a portion of the image L_{GRAY} shown in Fig. 5.27. The first operation is morphological closing performed with a disc-shaped structural element SE_K:

$$L_{KCLO} = (L_{GRAY} \oplus SE_K) \ominus SE_K \tag{5.27}$$

The function `SEK = strel('disk', 5)` was used for this purpose. The size of the structural element SE_K should match the size of the area where laser was applied—in this case it was 11×11 pixels. The image thus prepared can be subjected to further transformations—segmentation. Watershed segmentation was

Fig. 5.25 Image L_{GRAY} with automatically marked centres of laser effects

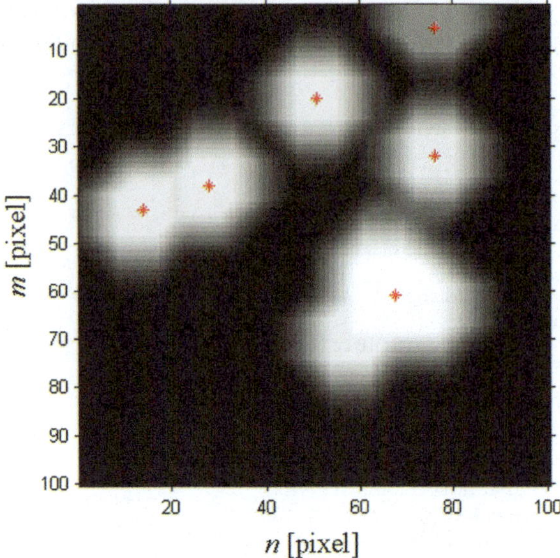

Fig. 5.26 Portion of the image L_{GRAY} covering the area immediately after laser application (as a 3D graph)

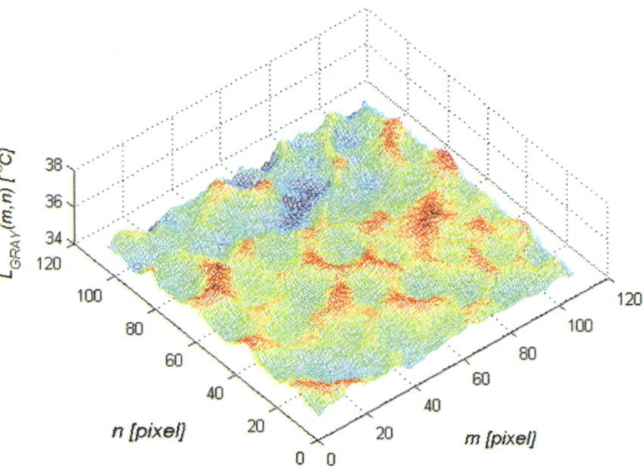

Fig. 5.27 Portion of the image L_{GRAY} covering the area immediately after laser application (as an image)

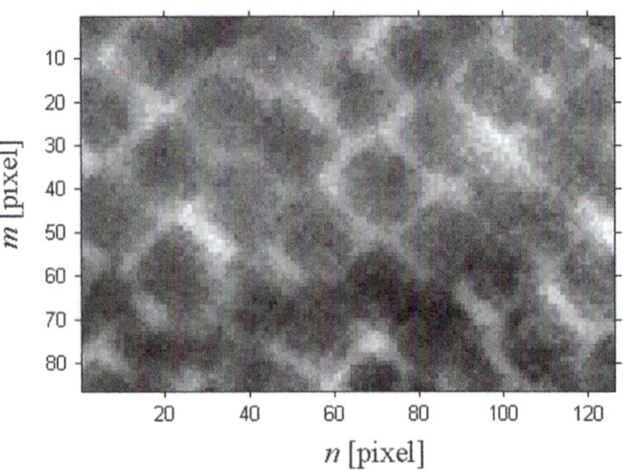

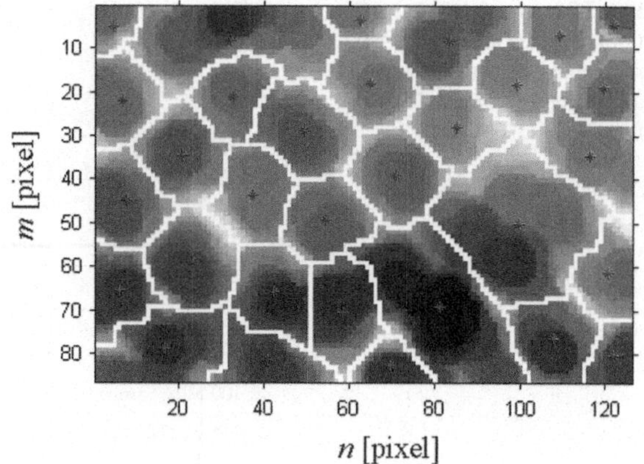

Fig. 5.28 Result, image L_{KCLO}, of morphological closing of the image L_{GRAY} with the structural element SE_K and results of watershed segmentation (*white edges*)

selected as it works very well in the case of the prepared image L_{KCLO}. The results of segmentation along with the image after morphological closing are shown in Fig. 5.28.

Because of the overlapping laser pulses, there frequently occur combined objects whose total area is approximately twice (or more times) larger than that of the others. Such situations can be observed in Figs. 5.29 and 5.30. Additionally, Fig. 5.30 shows in red the cut-off threshold p_{rg}:

$$p_{rg} = 1.5 \cdot \frac{\sum_{m=1}^{M} \sum_{n=1}^{N} \sum_{v=1}^{V} L_{BLAB}(m, n, v)}{V} \tag{5.28}$$

where:

$$L_{BLAB}(m, n, v) = \begin{cases} 1 & if \quad L_{LAB}(m, n) = v \\ 0 & other \end{cases} \tag{5.29}$$

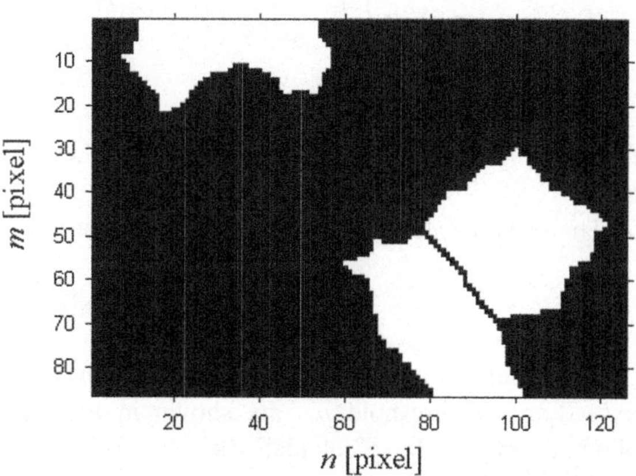

Fig. 5.29 Segmented areas of the image L_{KCLO} of surface area exceeding the value of p_{rg}

Fig. 5.30 Graph of changes in the surface area of the next segmented objects—*green*, and the cut-off threshold p_{rg}—*red*

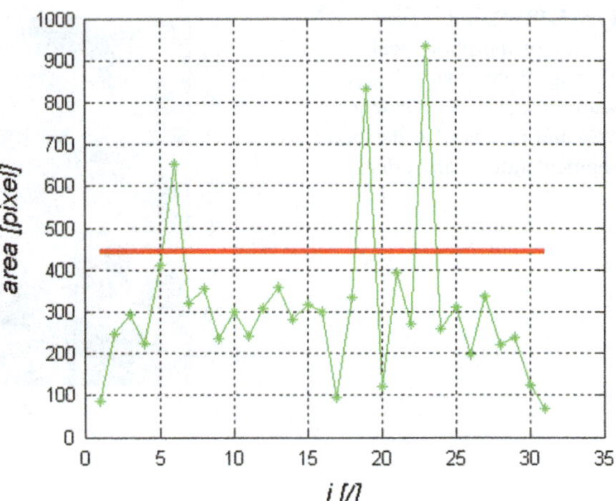

and L_{LAB} is the image with labels of individual segmented objects v.

The values of the surface area of objects v above the cut-off threshold p_{rg} will be analysed individually. Individual analysis applies to a situation where the value of the binarization threshold p_{rf} is chosen in such a way so as to minimize the value of the following criterion J_K:

$$J_K\left(p_{rf}\right) = 1 - \frac{\sum_{m=1}^{M}\sum_{n=1}^{N} L_{BK1}\left(m,n,p_{rf}\right)}{\sum_{m=1}^{M}\sum_{n=1}^{N} L_{BK2}\left(m,n,p_{rf}\right)} \tag{5.30}$$

where:

$$L_{BK1}\left(m,n,p_{rf}\right) = \begin{cases} 1 & if \\ 0 & other \end{cases} \quad \left(L_{KCLO}(m,n) > p_{rf}\right) \wedge \left(L_{BLAB}(m,n,v) = 1\right)$$

$$\tag{5.31}$$

$$L_{BK2}\left(m,n,p_{rf}\right) = \begin{cases} 1 & if \\ 0 & other \end{cases} \quad \left(L_{KCLO}(m,n) \leq p_{rf}\right) \wedge \left(L_{BLAB}(m,n,v) = 1\right)$$

$$\tag{5.32}$$

for L_{BLAB} including one object of the surface area exceeding the cut-off threshold p_{rg}.

The sought value of the binarization threshold p_{rf}^{*} is determined as:

$$p_{rf}^{*} = arg\left(min\left(J_K\left(p_{rf}\right)\right)\right) \tag{5.33}$$

The results of segmentation together with the additional division of objects exceeding the threshold p_{rg} are shown in Fig. 5.31. Below is the source code allowing for the above calculations:

```
LGRAY_=load('D:/core0004');
LGRAY=LGRAY_.core0004;
figure; imshow(LGRAY,[]); colormap('jet')
[LGRAY2]=imcrop(LGRAY,[189.51      161.51      125.98      85.98]);
LGRAY3=LGRAY2-273;
figure; mesh(LGRAY3); grid on;
xlabel('m [pixel]','FontSize',16,'FontAngle','Italic')
ylabel('n [pixel]','FontSize',16,'FontAngle','Italic')
zlabel('L_{GRAY}(m,n) [^OC]','FontSize',16,'FontAngle','Italic')
prg=35.6;
figure; imshow(LGRAY3,[])
SEK=strel('disk', 5)
LGRAY4=imclose(LGRAY3,SEK);
figure; imshow(LGRAY4,[])
LGRAY5=watershed(LGRAY4);
LGRAY6=LGRAY4; LGRAY6(LGRAY5==0)=max(LGRAY6(:));
figure; imshow(LGRAY6,[]); hold on
LLABEL=bwlabel(LGRAY5);
pam=[];
for i=1:max(LLABEL(:))
   Li=LLABEL==i;
   [xx,yy]=sr_ciez(Li);
   plot(xx,yy,'r*')
   pam(i,1:2)=[i, sum(sum(Li))];
end
figure; plot(pam(:,2),'-g*'); grid on; hold on
xlabel('i [/]','FontSize',16,'FontAngle','Italic')
ylabel('area [pixel]','FontSize',16,'FontAngle','Italic')
prm=median(pam(:,2))*1.5;
plot([1 pam(end,1)],[prm prm],'-r','LineWidth',3)
pam=[pam,pam(:,2)>prm];
pam(pam(:,3)==0,:)=[];
LGRAY7=zeros(size(LGRAY6));
LAB2=max(LLABEL(:))+1;
LLABEL2=LLABEL;
    if ~isempty(pam)
        for i=1:size(pam,1)
            Li=LLABEL==pam(i,1);
            LGRAY7=LGRAY7|Li;
            LGRAY8=LGRAY4.*Li;
            LGRAY8(Li==0)=[];
            pamr=[];
            for prf=min(LGRAY8):0.1:max(LGRAY8)
                pamr=[pamr;[prf,abs(1-sum(LGRAY8>prf)/sum(LGRAY8<=prf))]]
            end
            pamr_=sortrows(pamr,2);
            LLABEL2(((LGRAY4<pamr_(1,1)).*Li)==1)=LAB2;
            LAB2=LAB2+1;
        end
    end
    figure; imshow(LGRAY7)
    figure; imshow(LLABEL2,[]); colormap('jet'); colorbar; hold on
    pam=[];
    for i=1:max(LLABEL2(:))
        Li=LLABEL2==i;
        [xx,yy]=sr_ciez(Li);
        plot(xx,yy,'r*')
        pam(i,1:2)=[i, sum(sum(Li))];
    end
    figure; plot(pam(:,2),'-g*'); grid on; hold on
    xlabel('i [/]','FontSize',16,'FontAngle','Italic')
    ylabel('area [pixel]','FontSize',16,'FontAngle','Italic')
```

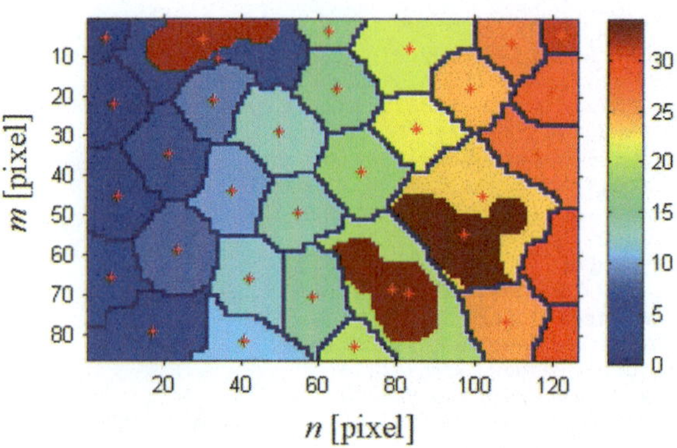

Fig. 5.31 Subsequent labels v assigned to individual objects in the modified image L_{LAB} with an additional division of three objects. Jet colour palette was adopted

In the next stage, the centres of gravity of individual objects are designated and their position is approximated with a square whose size corresponds to that of the area where laser was applied.

5.3 Segmentation of Some Selected Objects in Infrared Images

In general, it is not difficult to implement segmentation in infrared images. As also mentioned in previous chapters, simple binarization is sufficient when measurements are properly performed (no other warmer objects in the background). However, there are some thermal images in which, despite the time-consuming preparation of the patient for examination, it was not possible. These are cases in which the patient due to hospitalization must be against a warmer (or of comparable temperature) object, for example, newborns who need to stay in incubators or patients confined to bed. In these cases, segmentation of the object is much more complicated. What is more, it is more difficult than segmentation of typical objects in visible light. Sometimes the contour of the object is not clearly visible or completely invisible, and the object has both concave and convex irregular contour sections. There are also contour fragments which are difficult to determine even for a medical specialist. It should be emphasized that the cases which are analysed here are those for which it was assumed that there is only an infrared image (there is no equivalent image taken in visible light). Of course, it is possible to test the quality of the results obtained from the other known segmentation methods: region growing [142, 143], split and merge [144, 145], watershed segmentation [146, 147] or snake [148–152]. However, due to partially visible outline, the hybrid methods work best. One such method is the proposed method based on contour tracing.

Initially it was assumed that the method is semi-automatic (later this assumption was removed) [153, 154]. The operator must manually indicate the start point of the contour. Then, as a result of subsequent iterations of the algorithm, the next contour points are designated.

In the first stage, the point with coordinates $(m_i(1),n_i(1))$ is marked by the operator—see Fig. 5.32.

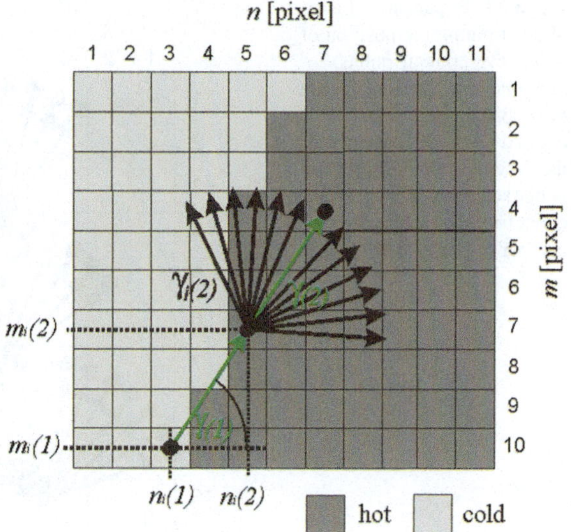

Fig. 5.32 Explanatory figure of the nomenclature adopted when determining the position of successive contour points $(m_i(1),n_i(1))$, $(m_i(2),n_i(2))$

Then the position of the next point $(m_i(2),n_i(2))$ is assumed. Because it is not known, there are two possibilities. The first one involves determination of the tangent to the contour edge already at this first stage. This method is deliberately not implemented here—at this point I encourage readers to implement it. The second method is the adoption of a fixed point shift $(m_i(2),n_i(2))$ with respect to the point $(m_i(1),n_i(1))$. This obviously yields false results for the first iterations but also simplifies the presented algorithm. In the next step, the angle $\gamma(g_i)$ is calculated, i.e.:

$$\gamma(g_i) = a\,cos\left(\frac{n_i(g_i - 1) - n_i(g_i)}{\sqrt{|n_i(g_i - 1) - n_i(g_i)|^2 + |m_i(g_i - 1) - m_i(g_i)|^2}}\right)$$
$$\cdot\,sign(m_i(g_i - 1) - m_i(g_i)) \tag{5.34}$$

for $g_i \in (2,G_i)$,

  where $sign$—returns the sign of the argument and

  G_i—is the total number of iterations adopted for testing $G_i = 500$.

The values for $g_i = 1$ are calculated in the first part of the algorithm at its very beginning. The acceptable ranges of changes in the angle $\gamma(g_i)$ are calculated in the subsequent stages. It is connected with the need to seek new contour points in a declared range. These in turn are associated with defining the angular range of search in relation to the previous angle. The range of $\pm\pi/2$ changed every 0.1 was adopted (see Fig. 5.33). Typically, when determining the position of a new point, $(m_i(3),n_i(3))$, $(m_i(4),n_i(4))$ etc., 32 possible positions are determined. A potential new position $\gamma_i(g_i,g_g) \in (\gamma(g_i) - \gamma_s,\gamma(g_i) + \gamma_s)$, where γ_s is a constant (parameter) set in the algorithm. For each angular value $\gamma_i(g_i,g_g)$, new positions of points m_2 and n_2 are calculated, i.e.:

Fig. 5.33 Explanatory figure of determining the position of successive contour points. Potential new locations for the next contour points are marked in *black*. Extension of the previous contour direction is marked in *blue*, while the correct new direction of the contour in *green*

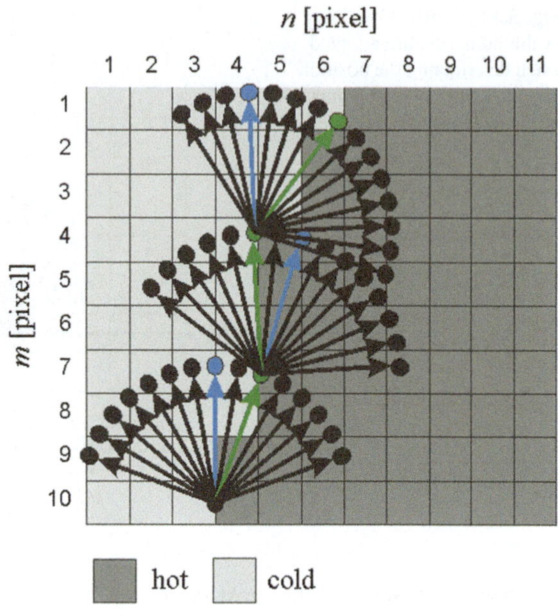

$$n_2\big(g_i, g_g\big) = \cos\big(\gamma_i\big(g_i, g_g\big)\big) + n_i(g_i - 1) \tag{5.35}$$

$$m_2\big(g_i, g_g\big) = \sin\big(\gamma_i\big(g_i, g_g\big)\big) + m_i(g_i - 1) \tag{5.36}$$

Then, for each angular value $\gamma_i(g_i, g_g)$, perpendicular lines passing through the point (m_2, n_2) are calculated, i.e.:

$$n_4\big(g_i, g_g, g_d\big) = \cos\Big(\gamma_i\big(g_i, g_g\big) + \frac{\pi}{2}\Big) \cdot d_{mn}(g_d) + n_2\big(g_i, g_g\big) \tag{5.37}$$

$$m_4\big(g_i, g_g, g_d\big) = \sin\Big(\gamma_i\big(g_i, g_g\big) + \frac{\pi}{2}\Big) \cdot d_{mn}(g_d) + m_2\big(g_i, g_g\big) \tag{5.38}$$

where $d_{mn}(g_d)$ is a parameter responsible for the total length of the line perpendicular to each potential contour point. Thus, this parameter ($d_{mn}(g_d)$) is responsible for the number of pixels included in the calculations. For each new potential position of the contour, the value of criterion $J_s(g_i, g_g)$ is calculated, i.e.:

$$J_s\big(g_i, g_g\big) = \Bigg| \frac{2}{G_d} \sum_{g_d=1}^{Gd/2} L_{GRAY}\big(m_4\big(g_i, g_g, g_d\big), n_4\big(g_i, g_g, g_d\big)\big)$$

$$- \frac{2}{G_d} \sum_{g_d=Gd/2}^{Gd} L_{GRAY}\big(m_4\big(g_i, g_g, g_d\big), n_4\big(g_i, g_g, g_d\big)\big) \Bigg| \tag{5.39}$$

Such a value of g_i—marked further as g_i^*—is sought for which:

$$g_{iL}^*(g_g) = arg\left(\min_{g_i}\left(J_s(g_i, g_g)\right)\right) \qquad (5.40)$$

when the sought object is inside (subscript Left—g_{iL}^*) and when the object is outside the marked starting point (subscript Right—g_{iR}^*):

$$g_{iR}^*(g_g) = arg\left(\max_{g_i}\left(J_s(g_i, g_g)\right)\right) \qquad (5.41)$$

The source code for carrying out the above calculations is presented below. In its first part, the coordinates of the point which the user clicked are downloaded, i.e.:

```
Lvar=LGRAY;
[n,m,klaw]=ginput(1);
```

Then, according to the adpted assumption, the next point is determined arbitrarily, shifted by +2 and −3 with respect to the point selected by the user.

```
ni=[n,n+2];
mi=[m,m-3];
```

In this step, calculations are made in accordance with the formula (5.34), i.e.:

```
gamma=acos(((ni(2)-ni(1))/sqrt(abs(ni(2)-ni(1))^2+
abs(mi(2)-mi(1))^2))*sign(mi(2)-mi(1));
```

In the next stages, a loop is formed which enables to carry out the calculations for the next iterations. Assuming the value of the parameter $d_{mn} \in (-4,4)$ pixels and 500 iterations:

```
line(ni,mi);
pam=50;
dmn=-4:4;
Ag=1;
for gi=2:500
    figure
    imshow(LGRAY,[],'notruesize')
    hold on
```

Subsequent calculations are directly related to the formulas (5.35) and (5.36):

```
gammas=pi*1/2;
    gammai=(gamma-gammas):0.1:(gamma+gammas);
    nv=Ag*cos(gammai)+ni(gi-1);
    mv=Ag*sin(gammai)+mi(gi-1);
    line(nv,mv,'Color','r');
    text(nv(1),mv(1),'*');
```

then according to the formulas (5.37) and (5.38):

```matlab
  Js=[];
  prze=[];
  for gg=1:length(nv)
      gamma=gammai(gg);
      nq=cos(gamma+pi/2)*dmn+nv(gg);
      mq=sin(gamma+pi/2)*dmn+mv(gg);
      nq=round(nq); mq=round(mq);
      nq(nq<1)=1; mq(mq<1)=1;
      [Mq,Nq]=size(LGRAY); nq(nq>Nq)=Nq;
  mq(mq>Mq)=Mq;
      line(nq,mq);
      for gii=1:length(mq);
      if ((nq(gii)==1)+(mq(gii)==1)+
       (nq(gii)==Nq)+(mq(gii)==Mq))>0
          prze(gg,gii)=pam(gii);
      else prze(gg,gii)=LGRAY(round(mq(gii)),
       round(nq(gii)));
      end
      end
      Js(gg)=abs(mean(prze(gg,round(length
      (nq)/2):length(nq))))-abs(mean(prze(gg,1:
      round(length(nq)/2)))));
      line([ni(gi-1),nv(gg)],[mi(gi-1),mv(gg)]);
      text(nq(1),mq(1),'*');
      pause
    end
 [~,gistar]=find(Js==min(Js));
```

Designation of the best new contour point according to the accepted criterion:

```matlab
gistar=gistar(1);
  gamma=gammai(gistar);
  nq=cos(gamma+pi/2)*dmn+nv(gg);
  mq=sin(gamma+pi/2)*dmn+mv(gg);
  nq=round(nq); mq=round(mq);
  nq(nq<1)=1; mq(mq<1)=1;
  [Mw,Nw]=size(LGRAY); nq(nq>Nw)=Nw; mq(mq>Mw)=Mw;
  for gii=1:length(mq);
  pam(gii)=LGRAY(round(mq(gii)),
   round(nq(gii)));
  end
  mi(gi)=mv(gistar);
  ni(gi)=nv(gistar);
  gamma=gammai(gistar);
  line([ni(gi-1),nv(gistar)],[mi(gi-1),mv(gistar)],'Color','r');
  ni(gi+1)=cos(gamma)+ni(gi);
  mi(gi+1)=sin(gamma)+mi(gi);
  line([ni(gi),ni(gi+1)],[mi(gi),mi(gi+1)],'Color','r');
  pause(0.1);
  gi
  Lvar(round(mi(gi)):round(mi(gi))+1,round
   (ni(gi)):round(ni(gi))+1)=min(LGRAY(:));
  imshow(Lvar,[],'notruesize'); colormap('jet'); colorbar
end
```

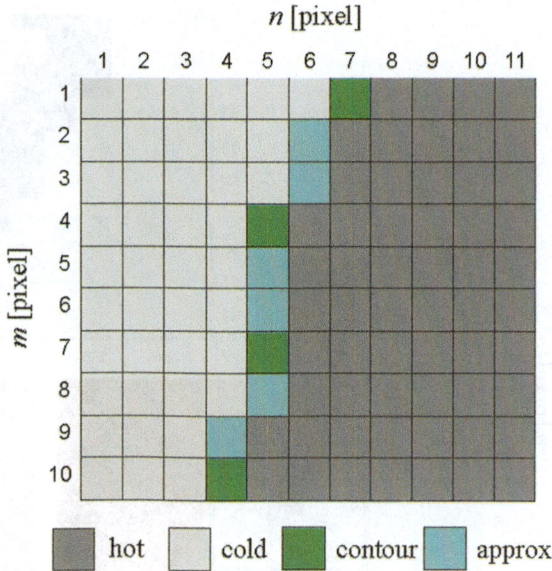

Fig. 5.34 Explanatory figure of the discrete position of pixels of the designated contour—according to Fig. 5.33. The pixels determined iteratively by seeking the position of subsequent points are marked in *green*. The pixels resulting from approximations between *green* pixels are marked in *blue*

One more parameter has been included in the above algorithm. It is directly responsible for the distance between the next sought points defined as variable A_g. For the results presented $A_g = 1$, unless otherwise stated. The results obtained are shown illustratively in Fig. 5.34, whereas the results for a sample infrared image are shown in Fig. 5.35.

The presented algorithm has very interesting properties resulting from its specific operation:

- d_{mn}—is the parameter responsible for the number of pixels on both sides of the contour that affects determination of new contour points. Its small values make the contour very sensitive to noise but it can also map an irregular shaped object. Large values mean that information in the area of contour detection is averaged from a larger area. Thus, in the case of detection of a small object contour and the proximity of another object, it may happen that the contour will start to approximate the other (wrong) object.

- γ_s is a parameter associated with the range of direction in which another contour pixel is sought. A small range of angular values will limit the arc radii which may be approximated. A large angular range increases proportionally computational complexity. So if there is any indication of the shape of the object contour arcs, it should be taken into account in the values of γ_s.

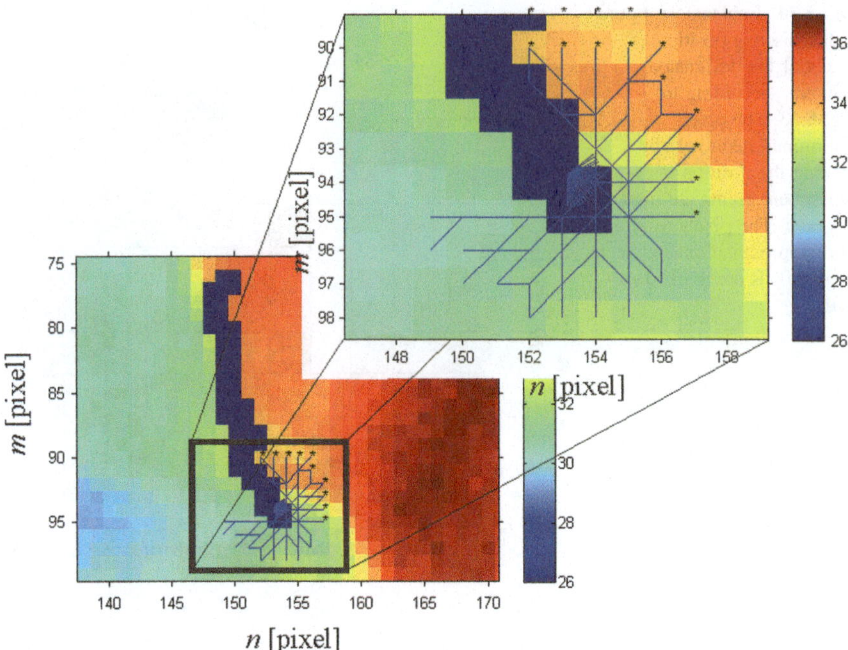

Fig. 5.35 Example of the results obtained according to the presented algorithm. The pixels that were taken into account when calculating the criterion $J(g_g)$ are marked with *thin blue lines*. The detected contour line of the adopted thickness of 2 pixels is marked in *navy blue*

- A_g is the third parameter. It is the distance between the subsequent contour points. For large distances between individual contour points, indirect, missing points, pixels must be interpolated. The advantage of large time distances between the determination of contour pixels is the speed of the algorithm operation. Determination of the position of each subsequent pixel results in much more accurate approximation of the contour and a proportional increase in computational complexity.
- G_i—the fourth parameter is the number of iterations. However, it will be further offset by the adopted stop criterion. The algorithm will be stopped when the same coordinates of the contour will overlap for the first time or when the maximum adopted number of iterations will be reached (as in the above algorithm—$G_i = 500$ iterations).

Different values of these four parameters are shown in Fig. 5.36 on the examples for the artificially generated image L_{GRAY}.

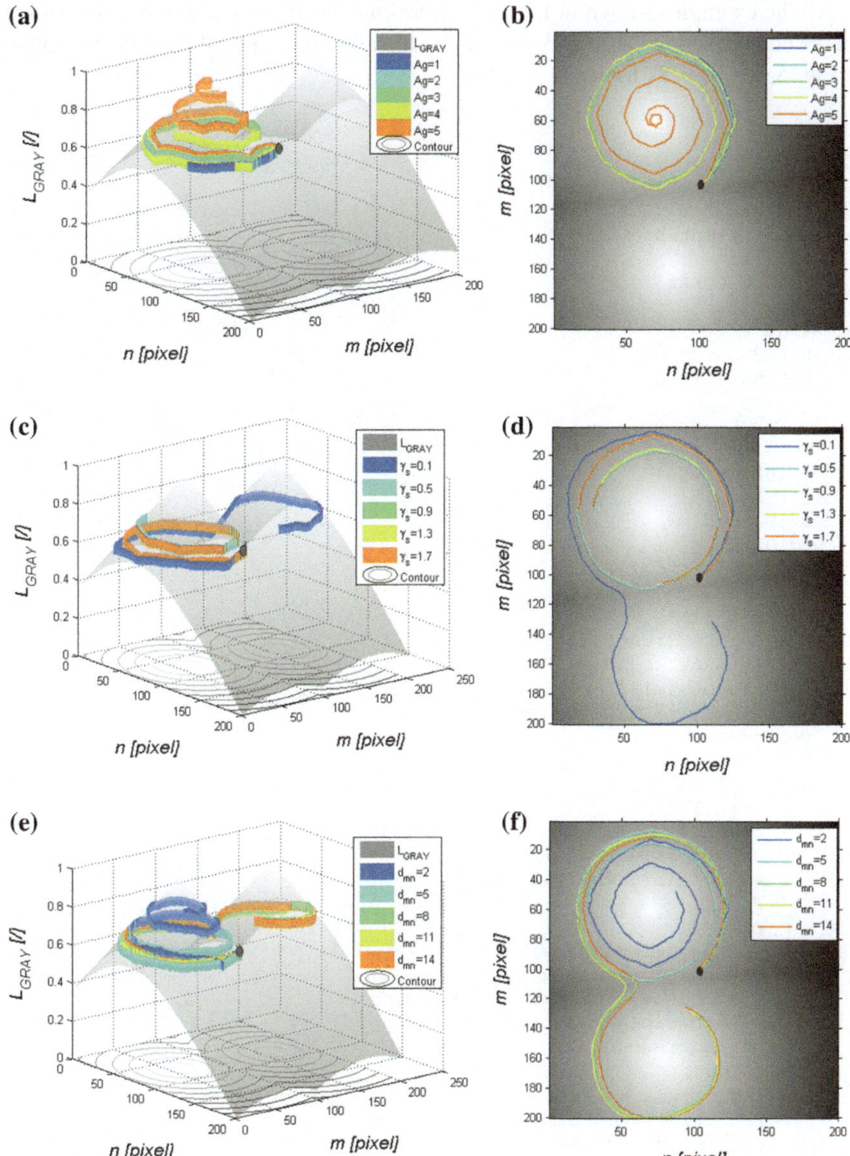

Fig. 5.36 Sample results for artificially generated image L_{GRAY} and different values of the three discussed algorithm parameters: **a**, **b** $\gamma_s = \pi/2$, $G_i = 100$, $A_g \in (1,5)$, $d_{mn} \in (-4,4)$; **c** and **d** $\gamma_s \in \{0.1, 0.5, 0.9, 1.3, 1.7\}$, $G_i = 500$, $A_g = 1$, $d_{mn} \in (-4,4)$; **e** and **f** $\gamma_s = \pi/2$, $G_i = 500$, $A_g = 1$, $d_{mn} \in (-2,2)$, $(-5,5)$, $(-8,8)$, $(-11,11)$, $(-14,14)$. The starting point indicated manually is marked with a *black dot*

All the examples shown in Fig. 5.36 were calculated using a modification of the source code described above. A sample modification providing the results presented in Fig. 5.36a, b is as follows:

```
LGRAY=zeros(200);
LGRAY(60,70)=1;
LGRAY(160,80)=1;
LGRAY=bwdist(LGRAY);
LGRAY=1-mat2gray(LGRAY);
figure
imshow(LGRAY,[],'notruesize');
hold on
figure
hObj2=surf(LGRAY);
set(hObj2,'FaceColor','interp',...
  'EdgeColor','none',...
  'FaceLighting','phong','FaceAlpha',0.3)
hold on
palet=jet(5);
for Ag=1:5;
m=102;
n=102;
zi=[];
ni=[n,n+2];
mi=[m,m-3];
zi(1)=LGRAY(round(mi(1)),round(ni(1)));
zi(2)=LGRAY(round(mi(2)),round(ni(2)));
gamma=acos((ni(2)-ni(1))/sqrt(abs(ni(2)-ni(1))^2+abs
(mi(2)-mi(1))^2))*sign(mi(2)-mi(1));
pam=[];
dmn=-4:4;
gammas=pi*1/2;
for gi=2:100
    gammai=(gamma-gammas):0.1:(gamma+gammas);
    nv=Ag*cos(gammai)+ni(gi-1);
    mv=Ag*sin(gammai)+mi(gi-1);
    Js=[];
    prze=[];
    for gg=1:length(nv)
        gamma=gammai(gg);
        nq=cos(gamma+pi/2)*dmn+nv(gg);
        mq=sin(gamma+pi/2)*dmn+mv(gg);
        nq=round(nq); mq=round(mq);
    nq(nq<1)=1; mq(mq<1)=1;
    [Mq,Nq]=size(LGRAY); nq(nq>Nq)=Nq; mq(mq>Mq)=Mq;
    for gii=1:length(mq);
    if ((nq(gii)==1)+(mq(gii)==1)+(nq(gii)==Nq)+
(mq(gii)==Mq))>0
        prze(gg,gii)=pam(gii);
    else
prze(gg,gii)=LGRAY(round(mq(gii)),round(nq(gii)));
    end
    end
```

```
Js(gg)=abs(mean(prze(gg,round(length(nq)/2):
length(nq))))-abs(mean(prze(gg,1:round(length(nq)/2)))));
    end
  [~,gistar]=find(Js==min(Js));
  gistar=gistar(1);
  gamma=gammai(gistar);
  nq=cos(gamma+pi/2)*dmn+nv(gg);
  mq=sin(gamma+pi/2)*dmn+mv(gg);
  nq=round(nq); mq=round(mq);
  nq(nq<1)=1; mq(mq<1)=1;
  [Mw,Nw]=size(LGRAY); nq(nq>Nw)=Nw; mq(mq>Mw)=Mw;
  for gii=1:length(mq);
      pam(gii)=LGRAY(round(mq(gii)),round(nq(gii)));
  end
  mi(gi)=mv(gistar);
  ni(gi)=nv(gistar);
  zi(gi)=LGRAY(round(mi(gi)),round(ni(gi)));
  gamma=gammai(gistar);
  ni(gi+1)=cos(gamma)+ni(gi);
  mi(gi+1)=sin(gamma)+mi(gi);
  zi(gi+1)=LGRAY(round(mi(gi+1)),round(ni(gi+1)));
end
tw=ones([1 length(mi)])*0;
verts = {[ni' mi' zi']};
hObj3=streamribbon(verts,{tw},0.05); hold on
set(hObj3,'EdgeColor','none','FaceLighting','phong',
'FaceColor',palet(Ag,:));
figure(1)
plot3(ni,mi,zi,'Color',palet(Ag,:),'LineWidth',2)
figure(2)
end
view(52,22)
colormap(gray)
contour(LGRAY)
legend('L_{GRAY}','Ag=1','Ag=2','Ag=3','Ag=4','Ag=5','Contour')
xlabel('n [pixel]','FontSize',16,'FontAngle','Italic')
ylabel('m [pixel]','FontSize',16,'FontAngle','Italic')
zlabel('L_{GRAY} [/]','FontSize',16,'FontAngle','Italic')
figure(1)
    legend('Ag=1','Ag=2','Ag=3','Ag=4','Ag=5')
    xlabel('n [pixel]','FontSize',16,'FontAngle','Italic')
    ylabel('m [pixel]','FontSize',16,'FontAngle','Italic')
```

An attentive reader will be able to independently carry out further modifications of the above source code to obtain the results shown in Fig. 5.36c–f.

The position of the starting point plays an important role in the algorithm. In the results shown in Fig. 5.36, this point is at a fixed location. When changing its position (for the other parameters), the results may vary. Some selected, extreme cases are shown in Fig. 5.37.

As expected and in accordance with the algorithm operation, the results obtained are closely related to the location of the starting point (Fig. 5.37). This problem can

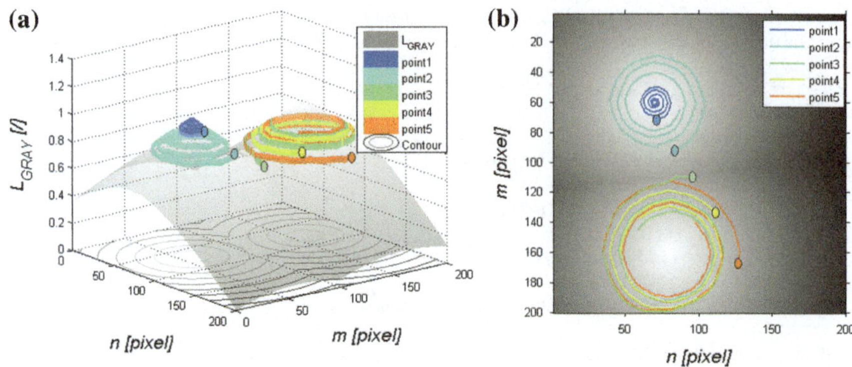

Fig. 5.37 Sample results for the artificially generated image L_{GRAY} and different indication of the starting point. The results were obtained for the following parameters: $\gamma_s = \pi/2$, $G_i = 500$, $A_g = 1$, $d_{mn} \in (-4,4)$. The starting points indicated manually are marked with *colourful points*

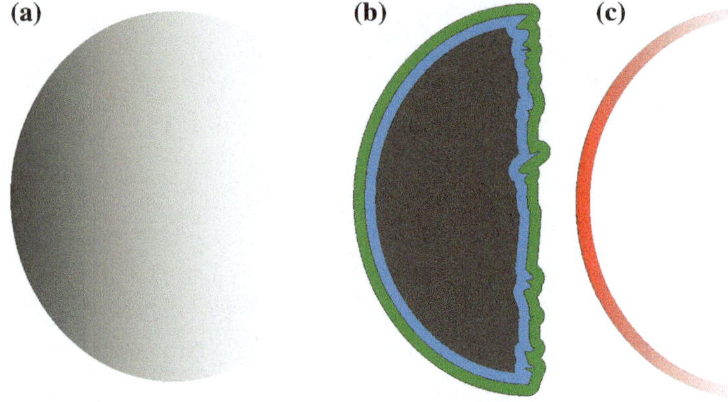

Fig. 5.38 Explanatory figure illustrating the operation of the algorithm portion responsible for the selection of the starting pixel (the object is shown in negative as *black*): **a** the object in the image L_{GRAY}; **b** the result of binarization and difference between the binary image and the result of erosion (*blue*) and the difference between the binary image and the result of dilation (*green*); **c** the contrast between the *blue* and *green* areas indicated with intensive *red* (more intensive, greater contrast)

be eliminated by using information about the object, and assuming that at least a few contour pixels (in extreme cases only one pixel) have good contrast against the background. This type of situation is illustrated in Fig. 5.38 and in the artificial image L_{GRAY} in Fig. 5.39.

The proposed method for automatic detection of the location of the starting point uses a matrix of the image L_{GRAY} and the image L_{BIN} resulting from binarization

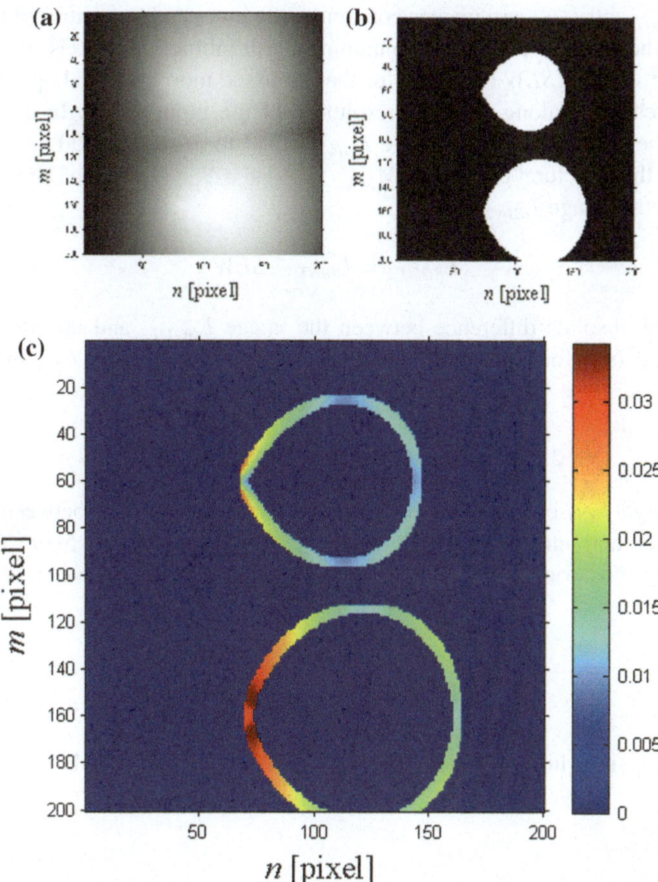

Fig. 5.39 Sample results of the analysis of the gradient of brightness changes in the area of the object contour: **a** image L_{GRAY}; **b** image L_{BIN}; and **c** L_{BOUT2} with artificial colour palette

(see formula 5.1). Different results are obtained depending on the adopted threshold p_r. However, this threshold (p_r) can be set on the basis of the adopted medical evidence, for example, the adopted threshold of temperature of subjects considered as healthy (36.6 [°C]). Then, in order to calculate the green and red areas shown in Fig. 5.38c, erosion and dilation must be calculated and a common portion with the image L_{BIN} must be determined, i.e. (analogously to formulas 5.3 and 5.4):

$$L_{BOUT} = ((L_{BIN} \oplus SEW) \veebar L_{BIN}) \cdot L_{GRAY} \tag{5.42}$$

$$L_{BINT} = ((L_{BIN} \ominus SEW) \veebar L_{BIN}) \cdot L_{GRAY} \tag{5.43}$$

where L_{BOUT} and L_{BINT} are images containing the inner and outer strip of the object visible in the binary image L_{BIN} containing the brightness of pixels of the image L_{GRAY}. The variable SEW is a mask for the discussed morphological operations and its size is changed along with the resolution of the input image. In the analysed case, for the resolution of the image L_{GRAY} equal to $M \times N = 200 \times 200$ pixels, the size of the structural element is $M_{SEW} \times N_{SEW} = 9 \times 9$ pixels. The next step is dilation in the image L_{BINT}, i.e.:

$$L_{BOUT2} = L_{BINT} \oplus SEW \tag{5.44}$$

Then the absolute difference between the image L_{BOUT2} and the image L_{BOUT} is calculated (see formula 5.42) in the region of the outer strip (see Fig. 5.38c), i.e.:

$$L_{GQ} = |L_{BOUT} - L_{BOUT2}| \cdot ((L_{BIN} \oplus SEW) \veebar L_{BIN}) \tag{5.45}$$

The image L_{GQ} contains information related to the contrast between the area of the outer and inner contour of the object—please see Fig. 5.39. Therefore, the starting position $(m_i(1), n_i(1))$ is the maximum pixel value in the image L_{GQ}, i.e.:

$$(m_i(1), n_i(1)) = arg\left(\max_{m,n} L_{GQ}(m, n) \right) \tag{5.46}$$

This analysis (which provided Fig. 5.39) was implemented in Matlab as follows:

```
LGRAY=zeros(200);
LGRAY(60,70)=1;
LGRAY(160,80)=1;
LGRAY=bwdist(LGRAY);
LGRAY=1-mat2gray(LGRAY);
LGRAY=LGRAY.*(ones(size(LGRAY,1),1)*linspace(0,1,size(LGRAY,2)));
figure
imshow(LGRAY,[],'notruesize');
hold on
LBIN=LGRAY>0.35;
figure
imshow(LBIN,'notruesize');
SEW=ones(9);
LBINT=xor(imerode(LBIN,SEW),LBIN).*LGRAY;
LBOUT=xor(imdilate(LBIN,SEW),LBIN).*LGRAY;
LBOUT2=imdilate(LBINT,SEW);
figure
```

```
imshow(LBOUT,'notruesize');
figure
imshow(LBOUT2,'notruesize');
figure
LGQ=abs(LBOUT-LBOUT2).*xor(imdilate(LBIN,SEW),LBIN);
imshow(LGQ,[],'notruesize');
colormap('jet'); colorbar
figure
imshow(LGRAY,[],'notruesize'); hold on
[mi,ni]=find(LGQ==max(LGQ(:)));
plot(ni(1),mi(1),'r*')
```

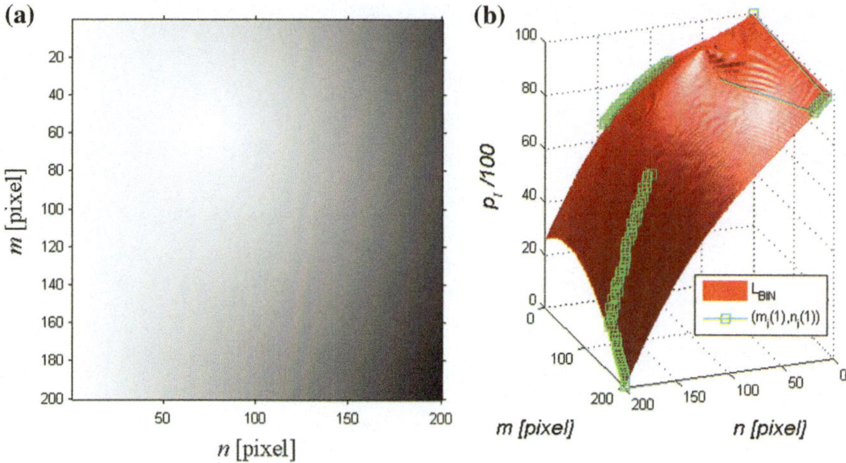

Fig. 5.40 Image L_{GRAY} and a sequence of binary images L_{BIN} for different values of the threshold p_r marked in *red* and the changing values of the starting point coordinates for the various thresholds p_r

In this way a complete algorithm is obtained that enables automatic analysis of objects taking into account determination of their contours. It works even when the information about the object is incomplete. The effect of the selection of the threshold p_r on the obtained results of determination of the starting point position is shown in Fig. 5.40.

Attention should be paid to two specific areas shown in the results in Fig. 5.40. The first region is associated with $p_r \in (0,0.6)$. The position of the starting point runs almost in a straight line relative to the centre of the object (Fig. 5.40). The second region is for $p_r \in [0.6,0.85)$, for which the starting point is on the other side of the object. The third region is for pr > 0.6, and due to the very small object (which is removed during the calculation of erosion—see formula 5.43), the position of starting points is calculated incorrectly. Below there is the source code allowing for these calculations:

```
LGRAY=zeros(200);
LGRAY(60,70)=1;
LGRAY=bwdist(LGRAY);
LGRAY=LGRAY.*(ones(size(LGRAY,1),1)*linspace(0,1,size(LGRAY,2)));
LGRAY=1-mat2gray(LGRAY);
figure; imshow(LGRAY)
SEW=ones(9);
ni=[]; mi=[]; zi=[]; LD=[];
for pr=0.01:0.01:1
    LBIN=LGRAY>pr;
    LD(1:size(LBIN,1),1:size(LBIN,2),round(pr*100))=LBIN;
    LBINT=xor(imerode(LBIN,SEW),LBIN).*LGRAY;
    LBOUT=xor(imdilate(LBIN,SEW),LBIN).*LGRAY;
    LBOUT2=imdilate(LBINT,SEW);
    LGQ=abs(LBOUT-LBOUT2).*xor(imdilate(LBIN,SEW),LBIN);
    [m,n]=find(LGQ==max(LGQ(:)));
    mi=[mi,m(1)];
    ni=[ni,n(1)];
    zi=[zi,pr*100];
end
figure;
LD=smooth3(LD,'box',5);
p = patch(isosurface(LD,.5), 'FaceColor', 'red',
'EdgeColor', 'none');
view(3);
daspect([1 1 .4])
camlight(40, 40);
camlight(-20,-10);
lighting gouraud
grid on
hold on
view(159,22)
plot3(ni,mi,zi,'-gs')
legend('L_{BIN}','(m_i(1),n_i(1))')
xlabel('n [pixel]','FontSize',16,'FontAngle','Italic')
ylabel('m [pixel]','FontSize',16,'FontAngle','Italic')
zlabel('p_r /100','FontSize',16,'FontAngle','Italic')
```

In the next step, the results obtained from the proposed algorithm were compared with the known segmentation methods. Therefore, tests were conducted for the artificial image L_{GRAY} and the real infrared image containing different types of objects such as: newborn, foot, hand, back of the patient. The results obtained for the artificial image L_{GRAY} are shown in Fig. 5.41 and for the real image (in this case the hand) in Fig. 5.42.

The first part of the source code responsible for the creation of an artificial image L_{GRAY} as well as binarization and watershed segmentation is shown below.

```
LGRAY=zeros(200);
LGRAY(60,70)=1;
LGRAY=bwdist(LGRAY);
LGRAY=LGRAY.*(ones(size(LGRAY,1),1)*linspace(0,1,size(LGRAY,2)));
LGRAY=1-mat2gray(LGRAY);
LGRAY(60:140,25:80)=0.8;
figure; mesh(LGRAY)
figure; imshow(LGRAY)
LBIN_TH=LGRAY<0.81;
figure; imshow(LBIN_TH)
LBIN_WA=watershed(LGRAY);
figure; imshow(LBIN_WA,[])
```

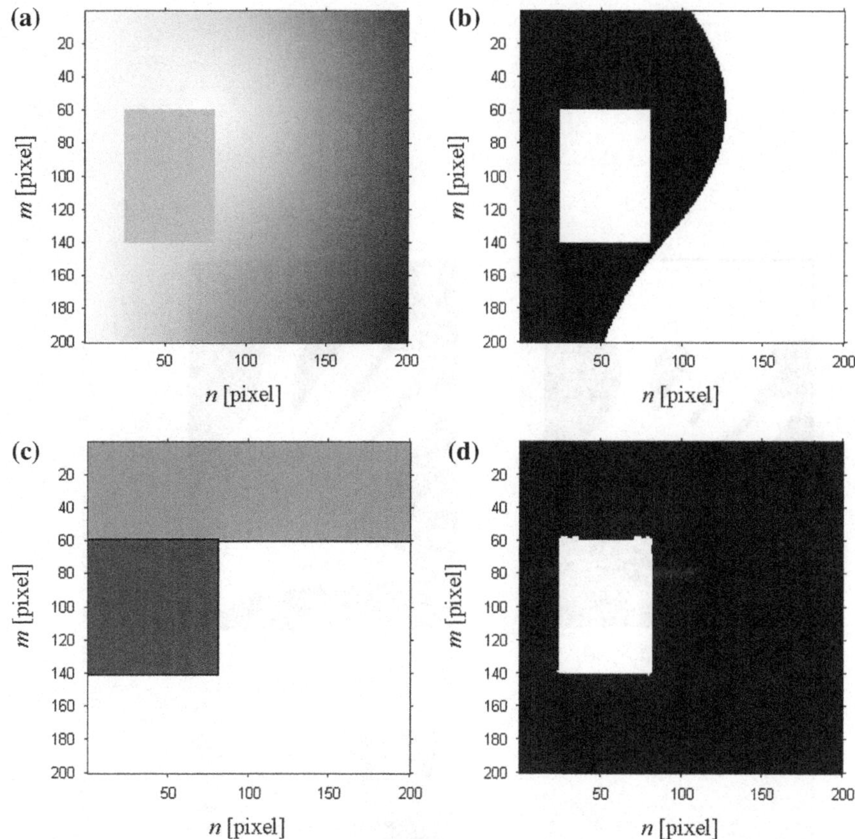

Fig. 5.41 Results of segmentation of the artificial image L_{GRAY} with different methods: **a** image L_{GRAY}; **b** binarization with the threshold of 0.81; **c** watershed method; **d** the method discussed in this book for $\gamma_s = \pi/2$, $G_i = 274$, $A_g = 1$, $d_{mn} \in (-4,4)$

Quantitative (using asterisks) comparison of the different selected types of segmentation is shown in Table 5.1.

The comparison of selected types of segmentation presented in Table 5.1 can be made by the readers themselves [155–158]. With the use of the functions `imrotate` and `imresize`, it is possible to analyse the effects of image rotation or resolution change on the results obtained. The result of this type of sample analysis of the impact of the image rotation by the angle $\phi \in (0°,40°)$ every $1°$ on the results obtained from the described algorithm is shown in Fig. 5.43.

As follows from the presented Table 5.1 and Fig. 5.43, the object analysis proposed in this book is competitive in relation to the active contour method. This is due to comparable ideas and methods of operation. The method of active contour by minimizing the external and internal energy of the object affects the position of the individual contour points. The shape of the starting contour (for the first iteration) is

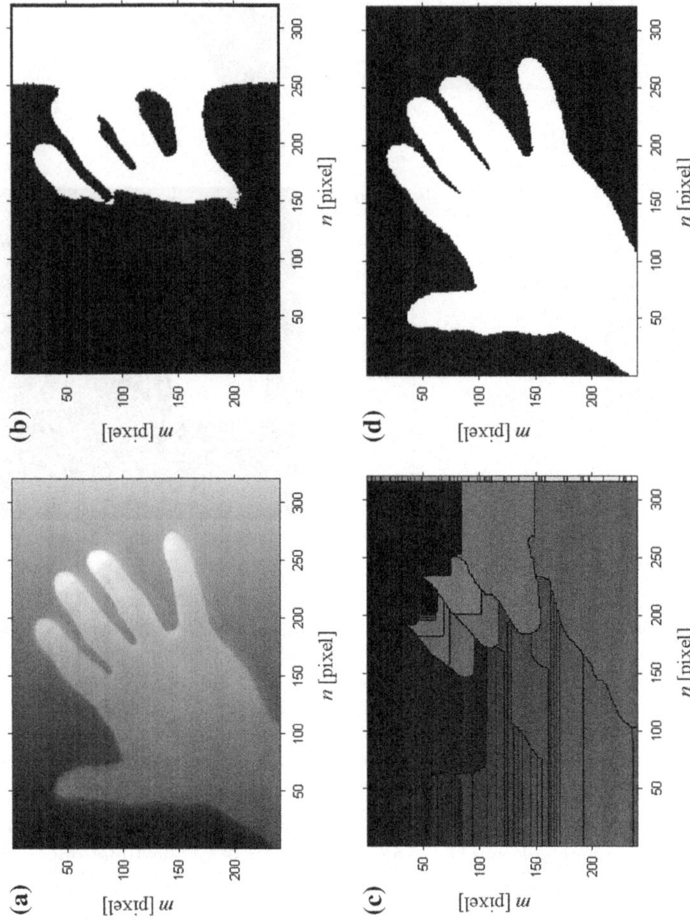

Fig. 5.42 Results of segmentation of the real image of the hand L_{GRAY} with different methods: **a** image L_{GRAY}; **b** binarization with the threshold of 26 °C; **c** watershed method; **d** the method discussed in this book for $\gamma_s = \pi/2$, $G_i = 1982$, $A_g = 1$, $d_{mn} \in (-4,4)$

Table 5.1 Comparison of characteristics of individual segmentation methods

Segmentation method	Computational complexity	Possibility of single object segmentation	Segmentation accuracy (difference in the surface area relative to the reference)	Sensitivity to image rotation
Binarization	*	No	**	No
Watersheds	**	No	*	No
Texture analysis	**	No	*	Yes
Methods based on artificial intelligence	***	Yes	***	Yes
Active contour method	**	Yes	***	No
Analysis proposed in the book	**	Yes	***	No

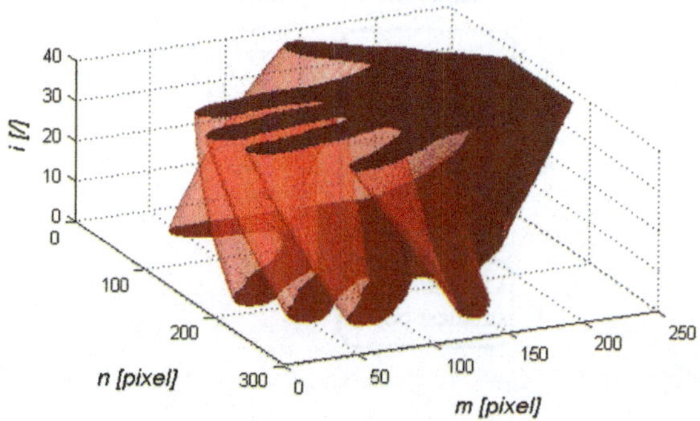

Fig. 5.43 Graph showing objects separated by means of the described algorithm for the image rotation by the angle $\phi \in (0°, 40°)$ every 1°. The values of the rotation angle ϕ are the same as the number of the next i image

generally rectangular [159–160]. The coordinates of the corners are usually the result of image pre-processing (rough localization of the object position) [161–168]. The angular range of seeking new contour points for the next iterations is usually limited (due to the location of adjacent points). This affects automatically the quality of the results obtained and the possibility of adapting the method to detect single lines and not isolated objects as, for example, in the balloon method (active contour) [169]. Consequently, the proposed method has the following characteristics:

- lower sensitivity to the selection of parameters,
- lower sensitivity to the initial position of the starting points,
- blurring an input image significantly affects the range of operation,

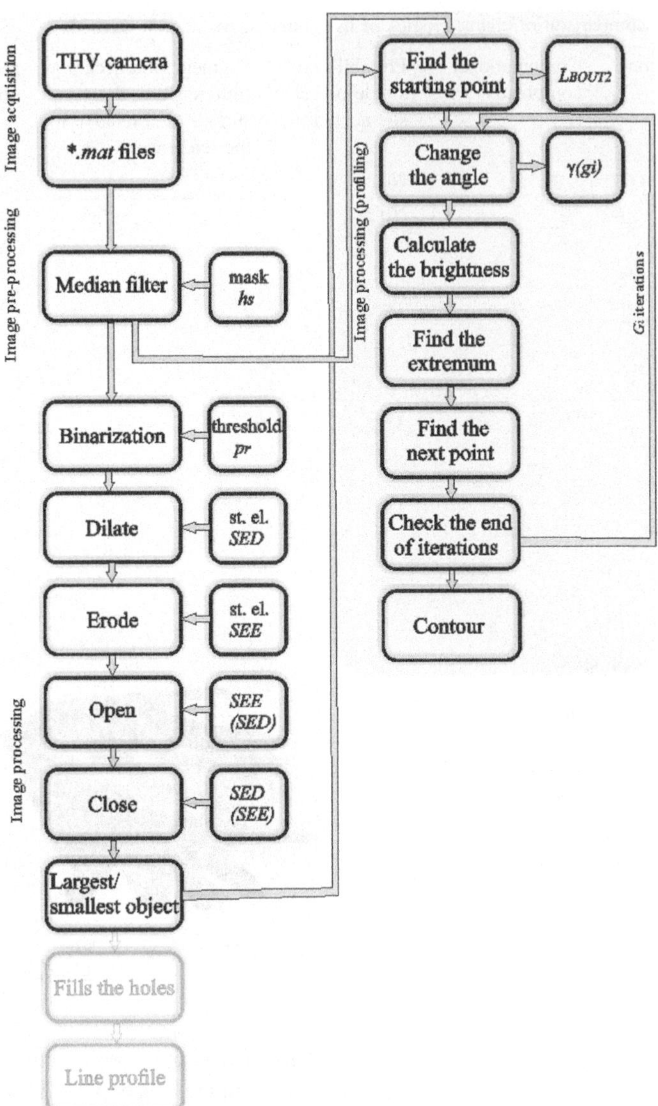

Fig. 5.44 Simplified block diagram of the automatic algorithm for the analysis of the object contour. The colours indicate four stages: image acquisition—*green*, image pre-processing—*blue*, image processing—*red* and image processing tailored to the calculations discussed in this subchapter—*red*. The significant results are marked with *green text*

- ability to detect concave shapes and other arbitrarily complex shapes,
- ability to automatically find the starting point so that the starting position of the points does not affect the results obtained,
- lower sensitivity to the position of the object relative to the image plane.

Therefore, the presented method can be also applicable to another type of images. These images can have unevenly illuminated objects. Virtually no segmentation methods can handle this type of images. A simplified block diagram of the whole algorithm is shown in Fig. 5.44 (compare with Fig. 5.22).

Chapter 6
Analysis of Image Sequences

Analysis of an image sequence concerns the analysis of images obtained during a single registration. Subsequent images of the same object form a video. However, due to the fact that in extreme cases there may be one image or two, this term ("video") is not used further. Instead, the term 'image sequence' is used. Each image in a sequence must be processed separately or jointly with the others. This processing involves the analysis presented in the previous chapters, i.e. image pre-processing, segmentation and separation of the object from the background. Analysis of the sequence will enable to examine temperature changes, position changes or time constants associated with the rate of temperature changes in the separated (segmented) image. This analysis applies directly to two aspects:

- digital stabilization of the position of the object in the scene and
- calculation of (e.g. minimum or maximum) temperature changes.

A detailed description of these two aspects is presented in the following subchapters.

6.1 Digital Image Stabilization

Digital image stabilization, which has been well known for many years [171–174], is a process allowing for stabilization of an object in the scene. There are various types of algorithms that make this type of analysis possible. Regardless of the algorithm type, it is necessary to acquire the characteristics of the object. These characteristics include geometric information about the object (such as the position of the edge, the position of the corners or local information on the pixel brightness).

© Springer International Publishing AG 2018
R. Koprowski, *Processing Medical Thermal Images*, Studies in Computational
Intelligence 716, DOI 10.1007/978-3-319-61340-6_6

One such method is the one proposed in the following description. This is a method using the correlation between the contours of subsequent objects. Its operation is based on the calculation of the correlation between i-th image and the first one. This correlation is calculated, in a simplified form, as xor operation (the resulting image $L_{BXOR}(m_s, n_s, i)$ and the indicator of the difference between the images is equal to $L_{BXOR2}(m, n, m_s, n_s, i)$):

$$L_{BXOR2}(m_s, n_s, i) = \sum_{n=1}^{N} \sum_{m=1}^{M} L_{BXOR}(m, n, m_s, n_s, i) \qquad (6.1)$$

$$L_{BXOR}(m, n, m_s, n_s, i) = L_{BIN}(m, n, i = 1) \veebar L_{BIN}(m + m_s, n + n_s, i \neq 1) \qquad (6.2)$$

for $i \in (1, I)$,

where m_s and n_s—are shifts of the image content resulting from the correlation search range, assuming that $m + m_s \in (1, M)$ and $n + n_s \in (1, N)$.

There is also another possible approach. Here stabilization takes place with respect to adjacent pairs of images (and not as in the above example, the first and the next image). Then the resulting image $L_{BXOR\_}(m_s, n_s, i)$ and the ratio of the difference between the images is equal to $L_{BXOR3}(m, n, m_s, n_s, i)$.

$$L_{BXOR3}(m_s, n_s, i) = \sum_{n=1}^{N} \sum_{m=1}^{M} L_{BXOR\_}(m, n, m_s, n_s, i) \qquad (6.3)$$

$$L_{BXOR}(m, n, m_s, n_s, i) = L_{BIN}(m, n, i) \veebar L_{BIN}(m + m_s, n + n_s, i + 1) \qquad (6.4)$$

for $i \in (1, I - 1)$.

In the first case, stabilization is performed relative to the first image or another image of the sequence specified by the operator. The advantage of this stabilization is obtainment of much better stabilization results if the angle of the camera position and visibility of the object in the scene do not change significantly. When the object is moving and the angle of the camera position changes, it is better to stabilize the image using the Eqs. (6.3) and (6.4). The values m_s and n_s, for which minimum $L_{BXOR}(m_s, n_s, i)$ or $L_{BXOR\_}(m_s, n_s, i)$ occur, are hereinafter referred to as m_s^* and n_s^*. It is further assumed that the range of variation (of image stabilization) is $m_s \in (-M/5, M/5)$, $n_s \in (-N/5, N/5)$, wherein $m + m_s \in (1, M)$ and $n + n_s \in (1, N)$. The above considerations assume that stabilization relates to only two coordinates—rows and columns of the image. In the general case there can occur two new variables—the third coordinate (the distance between the camera and the object) and object rotation in the scene. But adding one new variable means that optimization space increases by another dimension. It can be easily traced on the example of an artificially generated object (rectangle). In the first stage it was assumed that the object would move in the axis of

Fig. 6.1 Sequence of images $L_{GRAY}(m, n, i)$ without stabilization

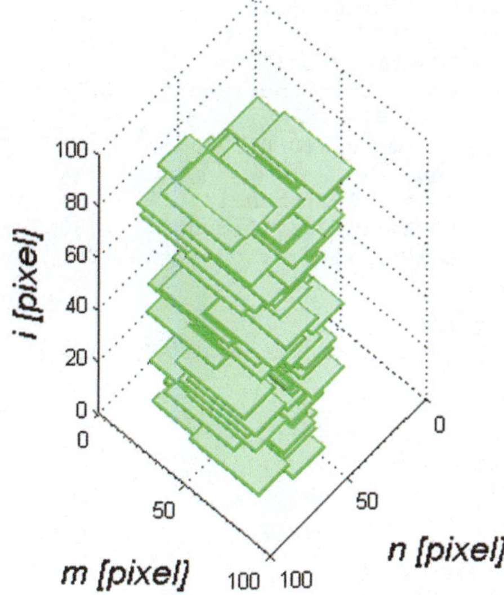

Fig. 6.2 Changes in the position of images $L_{GRAY}(m, n, i)$ for $i \in (1, I = 100)$ without stabilization (the values sought in stabilization are m_s^{**} and n_s^{**})

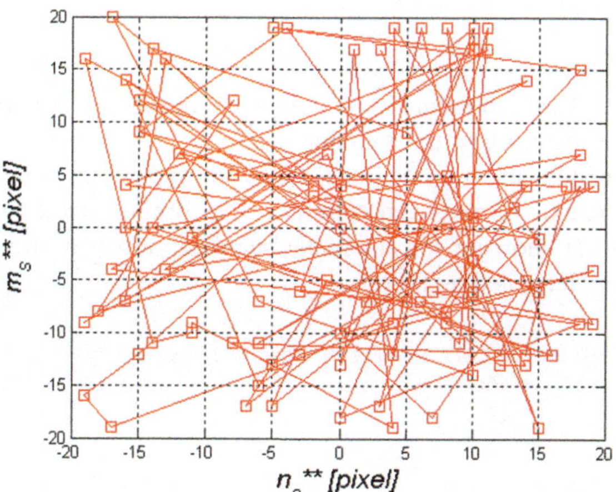

rows and columns at random. The proposed algorithm is designed to stabilize these shifts.

An artificial image resolution $M \times N = 100 \times 100$ pixels and $I = 100$ frames were assumed.

The source code enabling to create the graphs presented in Figs. 6.1 and 6.2 is shown.

```
LGRAY=[];msns=[];
for i=1:100;
    LGRAYi=zeros(100);
    msnsi=20-round(rand([1 2])*40);
    if i==1
        msnsi=[0 0];
    end
    msns(i,1:2)=msnsi;
    LGRAYi( (30:70)+msnsi(1),(50:70)+msnsi(2))=1;
    LGRAY(1:100,1:100,i)=LGRAYi;
end

figure;
p = patch(isosurface(LGRAY,.5), 'FaceColor', 'green', 'EdgeColor',
'none');
daspect([1 1 .4])
camlight(40, 40);
camlight(-20,-10);
lighting gouraud
grid on
hold on
view(132,64)
xlabel('n [pixel]','FontSize',16,'FontAngle','Italic')
ylabel('m [pixel]','FontSize',16,'FontAngle','Italic')
zlabel('i [pixel]','FontSize',16,'FontAngle','Italic')
figure; plot(msns(:,1),msns(:,2),'-rs'); grid on
xlabel('n_s** [pixel]','FontSize',16,'FontAngle','Italic')
ylabel('m_s** [pixel]','FontSize',16,'FontAngle','Italic')
```

All the presented functions have been used previously and are not commented
on here. Only the function `rand` in the loop `for` is noteworthy. It allows for
random selection (uniform distribution) of values ranging from ±20 pixels, both for
the axis of rows and columns. The formula written in the form of Eqs. (6.1) and
(6.2) allowing for automatic stabilization of the next image sequences has been
implemented as follows:

```
LGRAYS=LGRAY; mss=0; nss=0; time_=0;
for i=2:100
    tic
    LBXOR2=[];
    for ms=-20:20
        for ns=-20:20
            LBXOR2(ms+21,ns+21)=sum(sum(  xor( LGRAY(:,:,1) ,
transmn(LGRAY(:,:,i),ms,ns) )  ));
        end
    end
    [mssi,nssi]=find(LBXOR2==min(LBXOR2(:)));
    mss(i)=(mssi(1)-21);
    nss(i)=(nssi(1)-21);
    LGRAYS(:,:,i)=transmn(LGRAY(:,:,i),(mssi(1)-21),(nssi(1)-21));
    time_(i)=toc;
end
```

```
figure;
p = patch(isosurface(LGRAYS,.5), 'FaceColor', 'red', 'EdgeColor',
'none');
daspect([1 1 .4])
camlight(40, 40);
camlight(-20,-10);
lighting gouraud
grid on
hold on
view(132,64)
xlabel('n [pixel]','FontSize',16,'FontAngle','Italic')
ylabel('m [pixel]','FontSize',16,'FontAngle','Italic')
zlabel('i [pixel]','FontSize',16,'FontAngle','Italic')
figure; plot(msns(:,1)-mss',msns(:,2)-nss','-rs'); grid on
xlabel('n_s**-n_s* [pixel]','FontSize',16,'FontAngle','Italic')
ylabel('m_s**-m_s* [pixel]','FontSize',16,'FontAngle','Italic')
figure; plot(time_,'-gs'); grid on
xlabel('i','FontSize',16,'FontAngle','Italic')
ylabel('time [s]','FontSize',16,'FontAngle','Italic')
```

The results obtained after stabilization (image L_{GRAYS}) and differences between artificially specified shift (values m_s^{**} and n_s^{**}) and correction (values m_s^{*} and n_s^{*}) are shown in Figs. 6.3 and 6.4.

The graph of time needed to stabilize a single object is shown Fig. 6.5.

As shown in the graph in Fig. 6.5, mean time of stabilization of a single 2D image is 0.32 s (for a PC with Intel® Core i7 3.6 GHz, 64 GB RAM). The value for the first iteration is equal to 0 s as the calculations were not performed.

To compare the results of analysis time, image rotation in the range of λ $(-10°, 10°)$ has been added. The results of analysis time are shown in Fig. 6.6.

Fig. 6.3 Sequence of images $L_{GRAY}(m, n, i)$ after stabilization

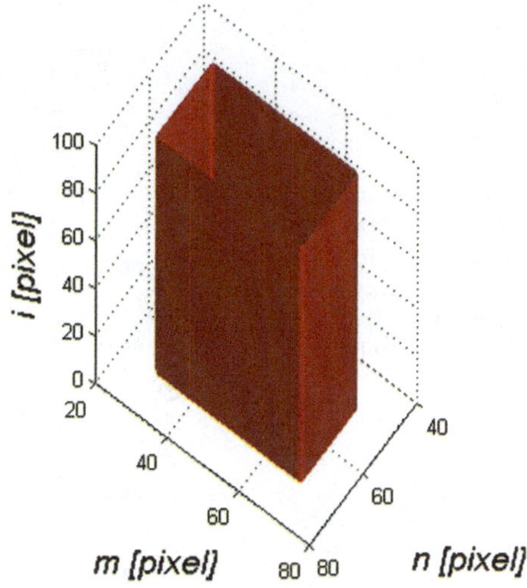

Fig. 6.4 Difference between artificially specified shift and (values m_s^{**} and n_s^{**}) and correction (values m_s^{*} and n_s^{*}) for $i \in (1, I = 100)$

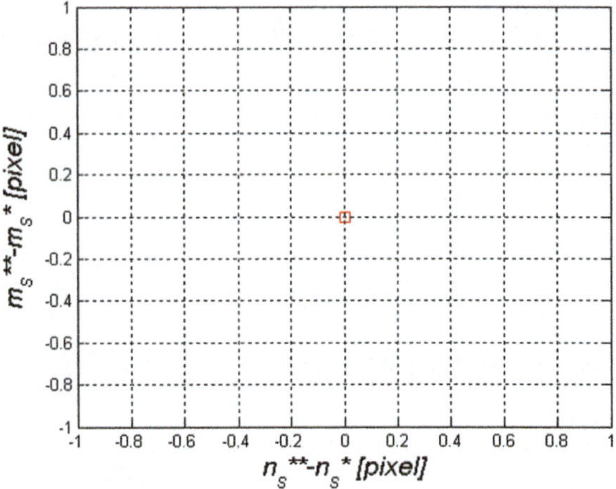

Fig. 6.5 Computation time for $i \in (1, I = 100)$ and two parameters—object shift in the rows and columns by ± 20 pixels

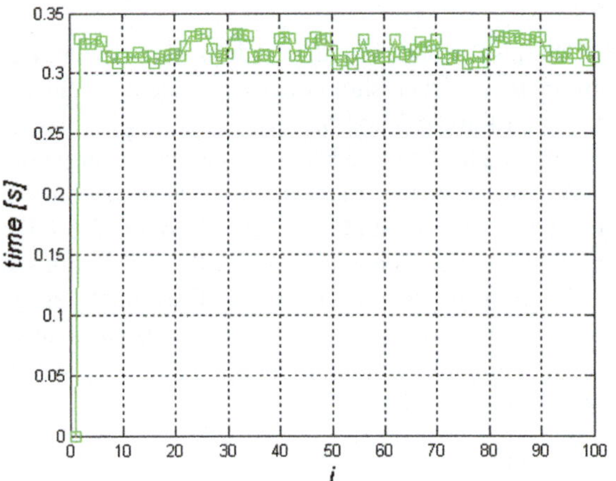

Fig. 6.6 Computation time for $i \in (1, I = 100)$ and three parameters—shift of the object in the rows and columns by ± 20 pixels and rotation by ± 10 [°]

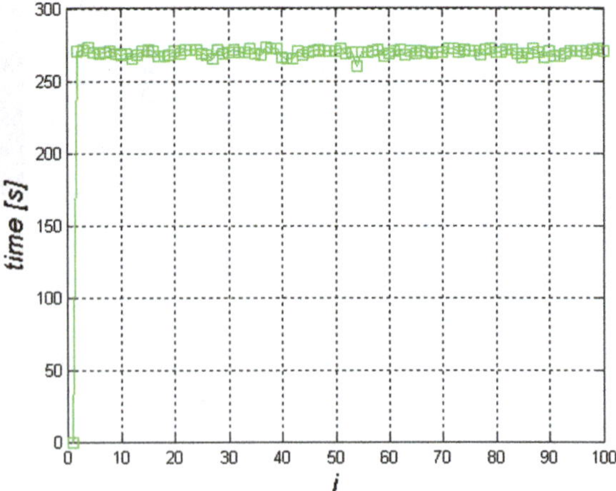

The stabilization time of a single image exceeds 260 s. It is more than 800 times slower than in the case when rotation is not analysed.

6.2   Serpentine Method of Infrared Image Analysis

A typical approach to infrared image analysis is the analysis of temperature changes. These changes are most often analysed at a particular point of the object or in the declared ROI. For a fixed object, this analysis of changes is typical and classical and generally does not pose major implementation difficulties. The situation is slightly different when the object is a living human organism and performs involuntary movements. In this case, a typical method involves movement stabilization and calculations. However, in most cases, these are oscillating movements. So the question remains whether it is possible, without recognizing the frequency of these movements in each individual case, to rearrange the sequence of images in such a way so as to stabilize the image. Or, whether it is possible to deliberately change the position of the object so that these changes would be useful from the point of view of further analysis—e.g. purposeful translation of the image allowing for the analysis of temperature changes around the selected point. Therefore, in this subchapter, serpentine method of image analysis is divided into two main areas:

- analysis of convolution of masks in a serpentine,
- analysis of image translation in a serpentine.

In the first case, the sequence of masks $h_{GA}(m_{GA}, n_{GA}, i_{GA}, \sigma_m, \sigma_n, \sigma_i)$ (where m_{GA}, n_{GA}, i_{GA} are subsequent normalized coordinates in a three-dimensional coordinate system and $\sigma_m, \sigma_n, \sigma_i$ are standard deviations of the mean in this system respectively), with the use of the Riesz pyramid [175–180] and Gaussian function for three dimensions, can be written as:

$$h_{GA}(m_{GA}, n_{GA}, i_{GA}, \sigma_m, \sigma_n, \sigma_i) = \frac{1}{\sigma_m \cdot \sigma_n \cdot \sigma_i \cdot (2 \cdot \pi)^{\frac{3}{2}}}$$
$$\cdot \exp\left(-\frac{m_{GA}^2}{2 \cdot \sigma_m^2} - \frac{n_{GA}^2}{2 \cdot \sigma_n^2} - \frac{i_{GA}^2}{2 \cdot \sigma_i^2} \right) \tag{6.5}$$

This approach assumes rotation of the mask h_{GA} in a flat, two-dimensional coordinate system by the angle θ multiplied by the coefficient k_w. So the new coordinates $(m_{GA\theta}, n_{GA\theta})$ after image rotation are equal to:

$$\begin{bmatrix} m_{GA\theta} \\ n_{GA\theta} \end{bmatrix} = \begin{bmatrix} \cos(\theta \cdot k_w) & -\sin(\theta \cdot k_w) \\ \sin(\theta \cdot k_w) & \cos(\theta \cdot k_w) \end{bmatrix} \cdot \begin{bmatrix} m_{GA} \\ n_{GA} \end{bmatrix} \qquad (6.6)$$

In practice, it may also be necessary to calculate derivatives in one, two or three dimensions of the mask h_{GA}, further in the source code referred to as dev. The source code allowing for the creation of a serpentine of masks h_{GA} is below:

```
hdevGA=[];
kw=18;
sigman=0.01;
sigmam=0.08;
sigmai=0.1;
devm=1;
devn=0;
devi=0;
M=100;N=100;I=440;
Ii=linspace(-0.5,0.5,I);
Iid=dergauss(Ii,sigmai,devi);
hObj = waitbar(0,'Please wait...');
for i=1:length(Iid);
    [nGA,mGA]=meshgrid(linspace(-0.5,0.5,M),linspace(-0.5,0.5,N));
    theta=i/I*pi*kw;
    nGAtheta=nGA.*cos(theta)+mGA.*sin(theta);
    mGAtheta=-nGA.*sin(theta)+mGA.*cos(theta);
    hdevGA(1:M,1:N,i)=dergauss(nGAtheta,sigman,devn).
   *dergauss (mGAtheta,sigmam,devm).*Iid(i);
    waitbar(i/length(Iid),hObj)
end
close(hObj)
figure
    [n,m,i]=meshgrid( 1:size(hdevGA,2) , 1:size(hdevGA,1) ,
    1:size(hdevGA,3));
    p1 = patch(isosurface(n,m,i,hdevGA>1,0.1),'FaceColor',
    [0 0 1 ],'EdgeColor','none');
    alpha(p1,0.9)
    p2 = patch(isosurface(n,m,i,hdevGA<-1,0.1),'FaceColor',
    [0 1 0 ],'EdgeColor','none');
    alpha(p2,0.9)
    view(41,24);
    camlight; camlight(-80,-10); lighting phong;
    grid on
    hold on
    axis([1 N 1 M 0 length(Iid)])
    axis image
    xlabel('m [pixel]','FontSize',14,'FontAngle','Italic')
    ylabel('n [pixel]','FontSize',14,'FontAngle','Italic')
    zlabel('i [pixel]','FontSize',14,'FontAngle','Italic')
```

and the function `dergauss` has the following source code:

```
function y = dergauss(x,sigma,w)
%    Copyright 2016 Robert Koprowski
%
%    Free only for readers Book
%    $Revision: 0.1 $  $Date: 2016/01/26 20:03:31 $
if w==0
    y = exp(-x.^2/(2*sigma^2)) / (sigma*sqrt(2*pi));
elseif w==1
    y =(-x./(sigma.^2)).*exp(-x.^2/(2.*sigma.^2)) ./
(sigma.*sqrt(2.*pi));
elseif w==2
    y =((x.^2-sigma.^2)./(sigma.^4)).*exp(-x.^2/(2.*sigma.^2)) ./
(sigma.*sqrt(2.*pi));
elseif w==3
    y =( (x.^3-3.*x.*sigma.^2) ./(sigma.^6)).
*exp(-x.^2/(2.*sigma.^2)) ./ (sigma.*sqrt(2.*pi));
elseif w==4
    y =( (x.^4-6.*x.^2.*sigma.^2+3.*sigma.^4)./(sigma.^8)).
*exp(-x.^2/(2.*sigma.^2)) ./ (sigma.*sqrt(2.*pi));
else
end
```

The use of the Riesz transform with the referenced source codes was discussed in detail in [181]. Therefore here only examples are provided without detailed discussion of their properties. One such example is the serpentine of masks for the first derivative calculated only along the axis of rows, for the resolution of $100 \times 100 \times 440$ points, $\sigma_m = 0.01$, $\sigma_n = 0.08$ and $\sigma_i = 0.1$, presented in Fig. 6.7.

Convolution of masks in a serpentine enables to detect sequential, repeated changes in the position of the object. The value of the coefficient k_w should be selected based on the expected rate of changes in the position of the object and the frame rate set during image acquisition.

The second option of the proposed serpentine approach is to analyse the object translation relative to the image plane. The idea of this method is shown in Fig. 6.8. As a result of affine transformations, the shift of the image content in the cyclic system is introduced. This approach will facilitate the analysis of temperature changes for the i-th sequence of infrared images and calculation of the parameters

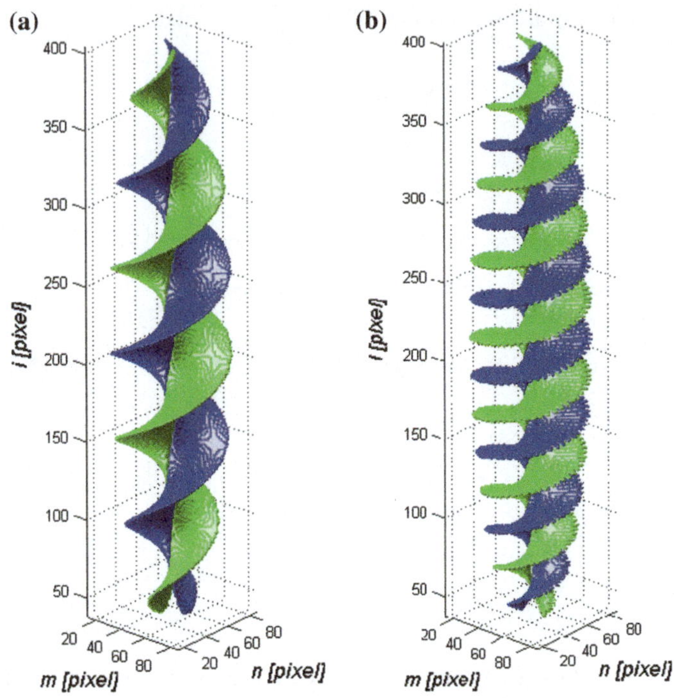

Fig. 6.7 Sequence of masks h_{devGA} for the serpentine method, resolution of $100 \times 100 \times 440$ points, $\sigma_m = 0.01$, $\sigma_n = 0.08$, $\sigma_i = 0.1$, and $k_w = 8$ image (**a**) and $k_w = 18$ image (**b**)

Fig. 6.8 Schematic diagram of the idea of serpentine analysis (translation) of the image viewed from the side and in the axis of changes

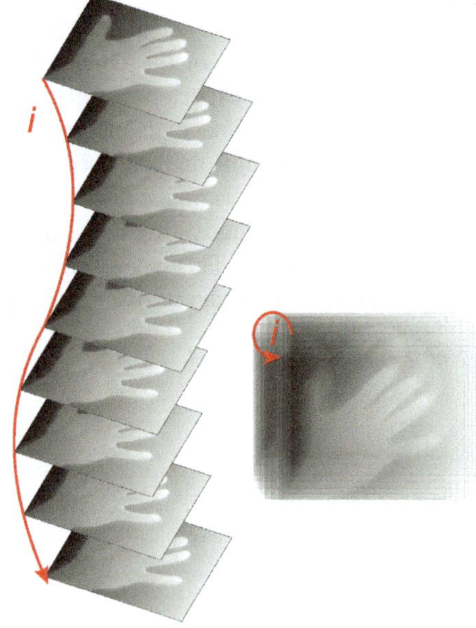

Fig. 6.9 Sample image sequence after shift stabilization for three values of temperature T (26, 29 and 30 °C)

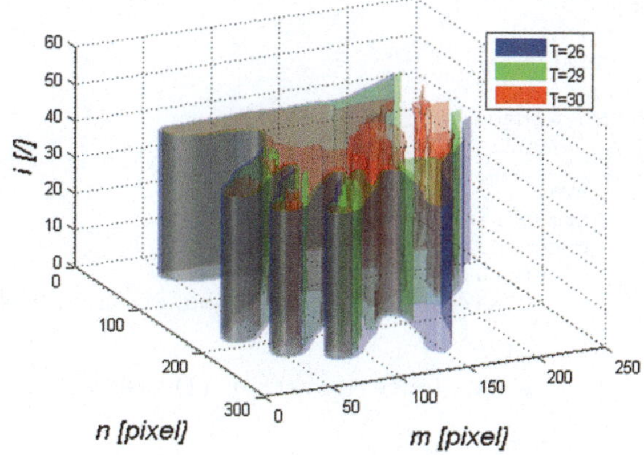

Fig. 6.10 Sample image sequence after affine shift of the object for three values of temperature T (26, 29 and 30 °C)

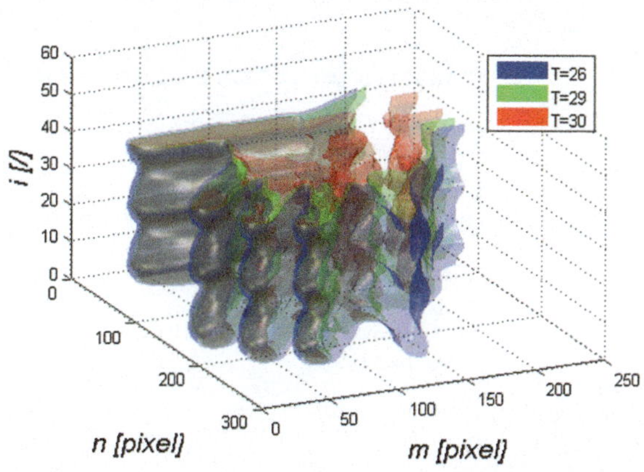

of a mathematical model. An example of an image sequence before and after the serpentine analysis is shown in Figs. 6.9 and 6.10.

The coefficient k_q was defined in analogy to the coefficient k_w in the previous method. This coefficient is responsible for the number of "twists" of the serpentine. In addition, the variable A_{mn} associated with the amplitude of changes (serpentine diameter) was introduced. The source code allowing for the implementation of such calculations is presented below.

```matlab
kq=0.2;
Amn=20;
LGRAYALL=[];
j=1;
LGRAY_=load('D:\reka3.mat');
for i=4333:(4333+40)
    eval(['LGRAY=LGRAY_.Frame',mat2str(i),';']);
    trm=abs(Amn*sin(kq*j));
    trn=abs(Amn*cos(kq*j));
    if trm>0
        LGRAY=[LGRAY((1+trm):end,:);ones([trm 1])*LGRAY(end,:)];
    end
    if trm<0
        LGRAY=[ones([abs(trm) 1])*LGRAY(1,:);LGRAY(1:end+trm,:)];
    end
    if trn>0
        LGRAY=[LGRAY(:,(1+trn):end),LGRAY(:,end)*ones([1 trn])];
    end
    if trn<0
        LGRAY=[LGRAY(:,1)*ones([1 abs(trn)]),LGRAY(:,1:end+trn)];
    end

        LGRAYALL(1:size(LGRAY,1),1:size(LGRAY,2),j)=LGRAY;
    j=j+1
end
figure;
LGRAYALL = smooth3(LGRAYALL,'box',5);
p1 = patch(isosurface(LGRAYALL,26), 'FaceColor', 'blue',
'EdgeColor', 'none'); hold on
p2 = patch(isosurface(LGRAYALL,29), 'FaceColor', 'green',
'EdgeColor', 'none');
p3 = patch(isosurface(LGRAYALL,30), 'FaceColor', 'red',
'EdgeColor', 'none');
alpha(p1,0.3)
alpha(p2,0.3)
alpha(p3,0.3)
view(65,18);
daspect([1 1 .4])
camlight(40, 40);
camlight(-20,-10);
lighting gouraud
grid on
hold on
xlabel('n [pixel]','FontSize',16,'FontAngle','Italic')
ylabel('m [pixel]','FontSize',16,'FontAngle','Italic')
zlabel('i [/]','FontSize',16,'FontAngle','Italic');
legend('T=26','T=29','T=30')
```

Due to the cyclical translation of the image sequence, it is possible to eliminate involuntary movements of the object. For this purpose, mean temperature in each area is analysed. An example of mean temperature change is shown in Figs. 6.11 and 6.12.

I leave to the readers implementation dedicated to a particular application and further considerations regarding the properties of the serpentine method.

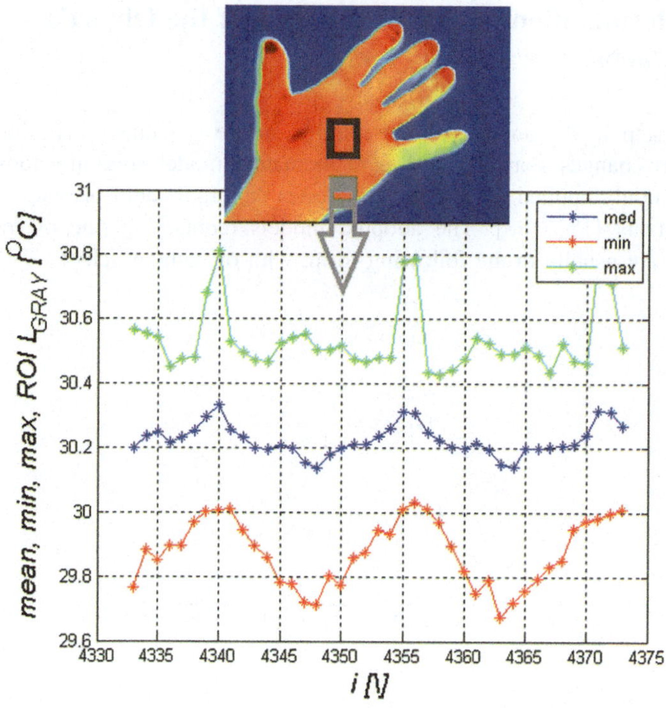

Fig. 6.11 Sample graph of changes in the mean, minimum and maximum temperature in the ROI in the serpentine approach for $i \in (4333, 4373)$

Fig. 6.12 Graph of changes in mean temperatures for different values of $k_q \in (0.01, 0.5)$, for $i \in (4333, 4373)$

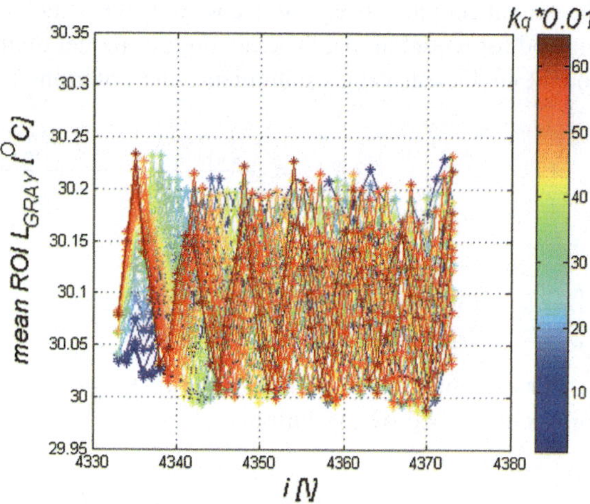

6.3 Determination of Time Constants of the Object's Response

The next step in the analysis of an infrared image sequence is the analysis of temperature changes. For this purpose, mathematical models used in automation are most commonly applied. They contain one (τ) or a maximum of two (τ_1 and τ_2) time constants [182–188]. The adopted models (containing one or more time constants) are usually in the following form, with or without delay:

$$G_B(s) = \frac{k_G}{1 + s \cdot \tau} \tag{6.7}$$

$$G_P(s) = \frac{k_G \cdot e^{-\tau_u s}}{1 + s \cdot \tau} \tag{6.8}$$

$$G_V(s) = \frac{k_G}{(1 + s \cdot \tau)^q} \tag{6.9}$$

$$G_A(s) = \frac{k_G}{(1 + s \cdot \tau_1) \cdot (1 + s \cdot \tau_2)} \tag{6.10}$$

where:

k_G gain coefficient.

Four different models were suggested above: $G_B(s)$ one-inert, $G_P(s)$ one-inert with delay, $G_V(s)$ multi-inert, $G_A(s)$ two-inert. There are many methods for finding the sought coefficients e.g. time constants. One such method is the Strejc method or method for model tuning. The method for model tuning is based on the selection of sought coefficient values minimizing the criterion J_G defined as:

$$J_G = \frac{\sum_{i=1}^{I} \left| T_{OUT}^{REAL}(i) - T_{OUT}^{MODEL}(i) \right|}{I} \tag{6.11}$$

where:

T_{OUT}^{REAL} response of the real object,
T_{OUT}^{MODEL} response of the object's model.

For example, for the object of the first order, a discrete spike response to force will be described by the formula:

$$T_{OUT}(i) = T_0 + (T_{max} - T_0) \cdot \left(1 - e^{-i/\tau}\right) \tag{6.12}$$

For the normalized case:

$$T_{OUT\_NORM}(i) = \frac{T_{OUT}(i) - T_0}{T_{max} - T_0} = 1 - e^{-i/\tau} \qquad (6.13)$$

Force (stimulation) in the practical case is implemented by means of halogen lamps arranged in sets. This type of stimulation, depending on the distance from the subject, can be adjusted with intensity (amplitude—the temperature value). The second case of stimulation is local cooling of the object, e.g. using Peltire's modules [189, 190, 191]. Then the object's return to the initial temperature (before cooling) is observed. A schematic graph of the response of a one-inert or two-inert system known from the literature (exactly from automation and control systems) and a block diagram of the idea of tuning parameters are shown in Figs. 6.13 and 6.14.

The sought object parameters were tuned using the function `fminsearch`. Proper operation of the method was initially evaluated for an artificial response of an inert object of the first order with added noise of uniform distribution in the range $(-1, 1)$. The following model parameters were adopted: $T_0 = 30$ [°C], $T_{max} = 35$ [°C] and $\tau = 15$ [s]. The corresponding source code allowing for calculations (of the sought object's parameters) is shown below:

```
tau=15;
T_0=30;
T_max=35;
t=0:100;
T_OUT_REAL=T_0+(T_max-T_0 )*(1-exp(-t/tau) );
figure; plot(t,T_OUT_REAL,'-g*'); grid on; hold on
T_OUT_REAL=T_OUT_REAL+(rand([size(t)])-0.5)*2;
plot(t,T_OUT_REAL,'-b*');
xlabel('i [\]','FontSize',16,'FontAngle','Italic')
ylabel('T_{OUT} [^OC]','FontSize',16,'FontAngle','Italic')
JG = @(x)sum(abs(T_OUT_REAL-(x(1)+(x(2)-x(1))*
(1-exp(-t/x(3)) )) ));

[x,fval] = fminsearch(JG,[10; 10; 10],...
    optimset('TolX',1e-28,'TolFun',1e-28,'MaxIter',10000000))
plot(t,x(1)+(x(2)-x(1))*(1-exp(-t/x(3))),'-r*');
legend('T_{OUT}^{REAL}','T_{OUT}^{REAL} +
noise','T_{OUT}^{MODEL}')
```

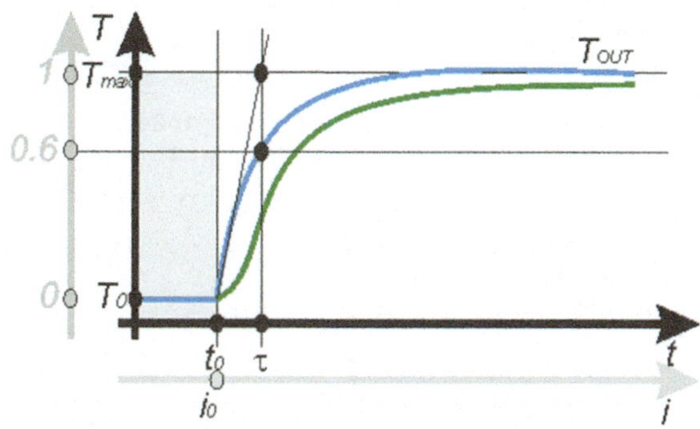

Fig. 6.13 Schematic graph of the response of a one-inert and two-inert object to a singular jump in temperature—continuous and discrete system as well as real and normalized temperature values. Stimulation in the form of object's cooling is marked with a *grey area*

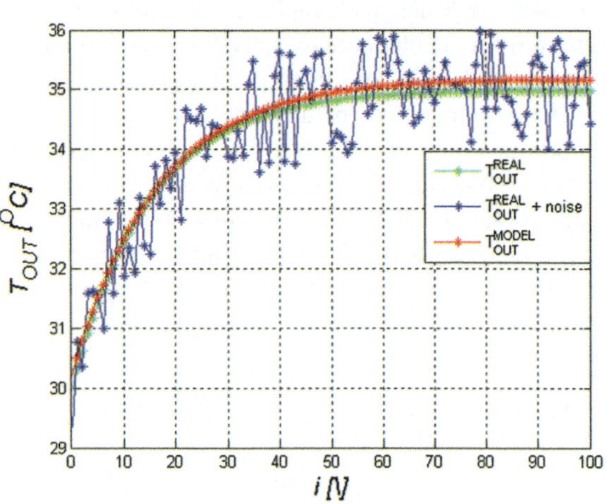

Fig. 6.14 General block diagram of the idea of tuning parameters (time constants τ_1 and τ_2) of the model for the criterion J_G

Fig. 6.15 Response of the object of the first order to a singular jump. The results from simulation and the object of automatically recognized parameters

The results obtained are $T_0 = 30.2$ [°C], $T_{max} = 35.1$ [°C] and $\tau = 16.4$ [s]. The graphs are shown in Fig. 6.15.

The presented results show that this method copes very well with finding the model parameters even in the presence of noise. The parameters found for the real object are shown in the form of images in Fig. 6.16. It should be added here that time constant values must be multiplied by the frame rate during acquisition, in this case 50. The corresponding source code allowing for these calculations is shown below.

```matlab
LGRAY_=load('D:\reka3.mat');
LGRAYALL=[];
j=1;
for i=4333:4514
    eval(['LGRAY=LGRAY_.Frame',mat2str(i),';']);
    LGRAYALL(1:size(LGRAY,1),1:size(LGRAY,2),j)=LGRAY;
    j=j+1;
end
L_T_0=zeros(size(LGRAY));
L_T_max=zeros(size(LGRAY));
L_tau=zeros(size(LGRAY));
t=(1:size(LGRAYALL,3))';
for m=1:size(LGRAYALL,1)
    for n=1:size(LGRAYALL,2)
```

```
        T_OUT_REAL(1:182,1,1)=LGRAYALL(m,n,:);
        JG = @(x)sum(abs(T_OUT_REAL-(x(1)+(x(2)-x(1))*
1-exp(-t/x(3)) )) ));
        [x,fval] = fminsearch(JG,[10; 10; 10],...
        optimset('TolX',1e-28,'TolFun',1e-28,'MaxIter',10000000));
        L_T_0(m,n)=x(1);
        L_T_max(m,n)=x(2);
        L_tau(m,n)=x(3);
    end
m
end
figure; imshow(L_T_0,[]); colormap('jet'); colorbar
figure; imshow(L_T_max,[]); colormap('jet'); colorbar
figure; imshow(L_tau,[]); colormap('jet'); colorbar
```

The resulting parameters of the time constant (or time constants for a two-inert object) represent new quality of information derived from the sequence of infrared images. The final step of both this as well as any of the above-mentioned algorithms should include the analysis of sensitivity *TPR* and specificity *SPC*, i.e.:

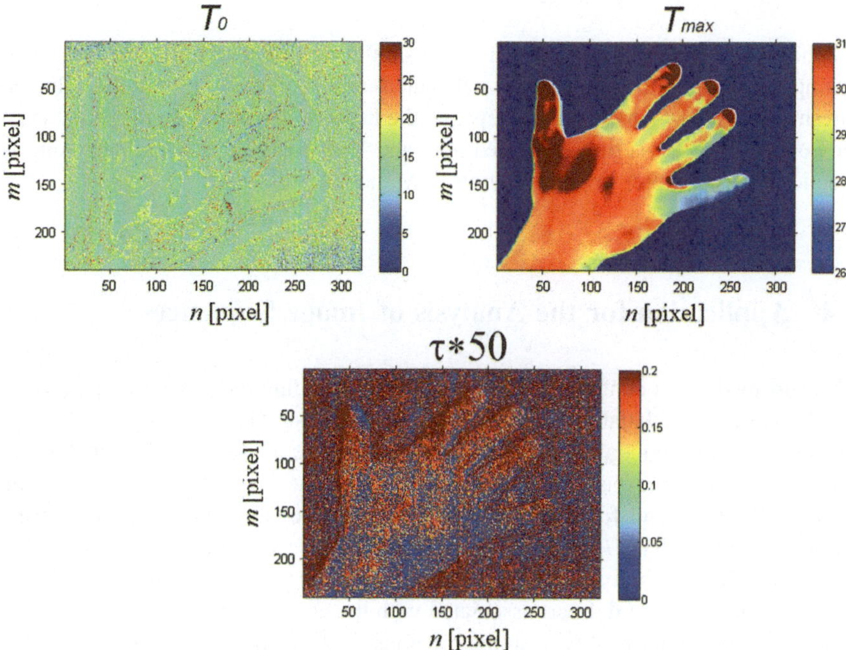

Fig. 6.16 Calculated parameters T_0, T_{max} and τ for the hand's temperature response (Fig. 6.11) after cooling presented in the form of images

$$TPR = \frac{TP}{TP + FN} \cdot 100\% \tag{6.14}$$

$$SPC = \frac{TN}{TN + FP} \cdot 100\% \tag{6.15}$$

where:

FP False Positive,
FN False Negative,
TP True Positive,
TN True Negative.

In addition, accuracy ACC is defined as follows:

$$ACC = \frac{TP + TN}{TP + TN + FP + FN} \cdot 100\% \tag{6.16}$$

These indicators (SPC, TPR and ACC) should be calculated, e.g. for the above example, taking into account the calculated time constants τ and the adopted threshold [192], which enables to separate the healthy subjects from the ill patients. However, this issue is only signalled here and I encourage readers to carry out its practical (very simple) implementation in Matlab.

6.4   Application for the Analysis of Image Sequences

This subchapter is a continuation of discussions regarding the proposed application —GUI created in Matlab. Chapter 5 proposes the basic GUI version designed for the analysis of individual images that do not constitute a sequence. The analysis of image sequences is associated with a significant modification of the file *THV_GUI_MAIN* and function *THV_GUI_sw*—further they are labelled with the index "2", i.e., *THV_GUI_MAIN2* and *THV_GUI2_sw* respectively. The m-file *THV_GUI_MAIN2* is divided in a similar manner as the file *THV_GUI_MAIN*. The first part contains the GUI part associated with the arrangement of buttons with the same or similar functionality as in the previous GUI (described in Chapter 5 of this monograph), i.e.:

- header and declaration of global variables:

```
%THV_GUI_MAIN2 graphical user interface
%    Copyright 2017 Robert Koprowski
%
%    Free only for readers book
global hObj LGRAY Lvar bin_var
```

- downloading the screen size in order to properly arrange individual windows:

```
scrsz = get(0,`ScreenSize`);
```

- creating artificial images in order to show them during GUI start-up:

```
LGRAY=rand(480,480); Lvar=LGRAY;
```

- path:

```
path=`D:/`;
cd(path);
```

- creating the main application window along with the basic menu buttons:

```
hObj(1)=figure('Name','THV_GUI_MAIN',...
               'NumberTitle','off');
col=get(hObj(1),'Color')*1.1; col(col>1)=1;
hObj(2)=uicontrol('Style', 'pushbutton',...
               'units','normalized',...
               'FontUnits','normalized',...
               'String', 'Open File(s)',...
               'Position', [0.001 0.95 0.15 0.05],...
               'Callback', 'THV_GUI2_sw(1)',...
               'BackgroundColor',col);
hObj(3)=uicontrol('Style', 'pushbutton',...
               'units','normalized',...
               'FontUnits','normalized',...
               'String', 'Save File(s)',...
               'Position', [0.001 0.9 0.15 0.05],...
               'Callback', 'THV_GUI2_sw(2)',...
               'BackgroundColor',col);
```

- buttons for changing the colour palette, median filter size, size of structural elements of dilation and erosion:

```matlab
hObj(4)=uicontrol('Style', 'popup',...
                  'units','normalized',...
                  'FontUnits','normalized',...
                  'String',
'autumn|bone|colorcube|cool|copper|gray|hot|hsv|jet|lines|
pink|prism|spring|summer|white|winter',...
                  'Position', [0.001 0.85 0.15 0.05],...
                  'Value',6,...
                  'Callback','THV_GUI2_sw(3)',...
                  'BackgroundColor',col);
hObj(5)=uicontrol('Style', 'popup',...
                  'units','normalized',...
                  'FontUnits','normalized',...
                  'String', 'None|Med 3x3|Med 5x5|Med 7x7|
Med 9x9|Med 11x11|Med 13x13|Med 15x15',...
                  'Position', [0.001 0.8 0.15 0.05],...
                  'Value',1,...
                  'Callback','THV_GUI2_sw(3)',...
                  'BackgroundColor',col);
bin_var=(0:0.1:1)';
hObj(6)=uicontrol('Style', 'popup',...
                  'units','normalized',...
                  'FontUnits','normalized',...
                  'String', {'None';num2str(bin_var)},...
                  'Position', [0.001 0.75 0.15 0.05],...
                  'Value',1,...
                  'Callback','THV_GUI2_sw(3)',...
                  'BackgroundColor',col);
hObj(7)=uicontrol('Style', 'popup',...
                  'units','normalized',...
                  'FontUnits','normalized',...
                  'String', 'None|Dil 3x3|Dil 5x5|Dil 7x7|
Dil 9x9|Dil 11x11|Dil 13x13|Dil 15x15',...
                  'Position', [0.001 0.7 0.15 0.05],...
                  'Value',1,...
                  'Callback','THV_GUI2_sw(3)',...
                  'BackgroundColor',col);
hObj(8)=uicontrol('Style', 'popup',...
                  'units','normalized',...
                  'FontUnits','normalized',...
                  'String', 'None|Ero 3x3|Ero 5x5|Ero 7x7|
Ero 9x9|Ero 11x11|Ero 13x13|Ero 15x15|',...
                  'Position', [0.001 0.65 0.15 0.05],...
                  'Value',1,...
                  'Callback','THV_GUI2_sw(3)',...
                  'BackgroundColor',col);
```

- buttons for changing the size of the structural elements of opening and closing:

```
hObj(9)=uicontrol('Style', 'popup',...
                  'units','normalized',...
                  'FontUnits','normalized',...
                  'String', 'None|Ope 3x3|Ope 5x5|Ope 7x7|
Ope 9x9|Ope 11x11|Ope 13x13|Ope 15x15|',...
                  'Position', [0.001 0.6 0.15 0.05],...
                  'Value',1,...
                  'Callback','THV_GUI2_sw(3)',...
                  'BackgroundColor',col);
hObj(10)=uicontrol('Style', 'popup',...
                  'units','normalized',...
                  'FontUnits','normalized',...
                  'String', 'None|Clo 3x3|Clo 5x5|Clo 7x7|
Clo 9x9|Clo 11x11|Clo 13x13|Clo 15x15|',...
                  'Position', [0.001 0.55 0.15 0.05],...
                  'Value',1,...
                  'Callback','THV_GUI2_sw(3)',...
                  'BackgroundColor',col);
```

- buttons for choosing one, two or three largest obejcts:

```
hObj(11)=uicontrol('Style', 'popup',...
                  'units','normalized',...
                  'FontUnits','normalized',...
                  'String', 'All|1Max Obj|2Max Obj|3Max Obj|
Min Obj|',...
                  'Position', [0.001 0.5 0.15 0.05],...
                  'Value',1,...
                  'Callback','THV_GUI2_sw(3)',...
                  'BackgroundColor',col);
```

- button for filling an object if it has holes:

```
hObj(12)=uicontrol('Style', 'popup',...
                  'units','normalized',...
                  'FontUnits','normalized',...
                  'String', 'None|Fill',...
                  'Position', [0.001 0.45 0.15 0.05],...
                  'Value',1,...
                  'Callback','THV_GUI2_sw(3)',...
                  'BackgroundColor',col);
hObj(13)=axes('Parent',hObj(1),...
                  'units','normalized',...
                  'Position', [0.3 0.2 0.52 0.72],'Visible','off');
```

- image display and setting the working area depending on its resolution:

```
hObj(14)=imshow(LGRAY,'DisplayRange',
[0 1],'InitialMagnification','fit','Parent',hObj(13));
hold on
axis([0 size(LGRAY,2) 0 size(LGRAY,1)]);
```

- showing two sliders for moving the horizontal and vertical line of cross-section:

```
hObj(15)=uicontrol('Style', 'slider',...
                   'units','normalized',....
                   'FontUnits','normalized',...
                   'BackgroundColor',col,...
                   'Position', [0.2 0.15 0.02 0.82],..
          'Min',1,'Max',480,'Value',1,...
                   'SliderStep',[1/100 1/100],...
                   'Callback', 'THV_GUI2_sw(4)');
hObj(16)=uicontrol('Style', 'slider',...
                   'units','normalized',....
                   'FontUnits','normalized',...
                   'BackgroundColor',col,...
                   'Position', [0.2 0.11 0.62 0.03],...
                   'Min',1,'Max',480,'Value',1,...
                   'SliderStep',[1/100 1/100],...
                   'Callback', 'THV_GUI2_sw(4)');
```

- showing the graph of horizontal and vertical brightness (temperature) changes:

```
hObj(17)=plot(sum(LGRAY,1),'-r*','LineWidth',2,'Parent',
hObj(13)); hold on; grid on
hObj(18)=plot(sum(LGRAY,2),'-g*','LineWidth',2,'Parent',hObj(13));
```

- showing the text related to the number of frames of an image sequence, date, time, and measurement parameters:

```
hObj(19)=uicontrol('Style', 'text',...
                   'units','normalized',...
                   'String', {'FileName = -','Frame = -',
 'Date = -','Time = -','ObjectParam = -'},...
                   'Position', [0.85 0.45 0.15 0.54],...
                   'BackgroundColor',col,...
                   'Parent',hObj(1));
```

- field for marking the possibility of viewing the original image (without processing) or the image after final processing:

```
hObj(20)=uicontrol('Style', 'checkbox',...
                   'units','normalized',...
                   'FontUnits','normalized',...
                   'String', 'Oryginal on/off',...
                   'Position', [0.001 0.2 0.15 0.05],...
                   'Value',0,...
                   'Callback','THV_GUI2_sw(3)',...
                   'BackgroundColor',col);
```

- field for marking the possibility of horizontal or vertical cross-section display or its lack:

```
hObj(21)=uicontrol('Style', 'checkbox',...
                   'units','normalized',...
                   'FontUnits','normalized',...
                   'String', 'Profile on/off',...
                   'Position', [0.001 0.15 0.15 0.05],...
                   'Value',1,...
                   'Callback','THV_GUI2_sw(5)',...
                   'BackgroundColor',col);
```

- field for marking the possibility of automatic data saving to Excel:

```
hObj(22)=uicontrol('Style', 'checkbox',...
                   'units','normalized',...
                   'FontUnits','normalized',...
                   'String', 'Excel write on/off',...
                   'Position', [0.001 0.10 0.15 0.05],...
                   'Value',1,...
                   'Callback','THV_GUI2_sw(6)',...
                   'BackgroundColor',col);
```

- new window for showing changes in the temperature on a 2D graph:

```
hObj(23)=figure('Name','THV_PROFILE_X_Y',...
                'NumberTitle','off');
hObj(24)=axes('Parent',hObj(23),...
              'units','normalized',...
              'Position', [0.1 0.1 0.85 0.85]);
hObj(25)=plot(sum(LGRAY,1),'-r*','LineWidth',2,
'Parent',hObj(24)); hold on; grid on
hObj(26)=plot(sum(LGRAY,2),'-g*','LineWidth',2,
'Parent',hObj(24));
                   xlabel('m,n [pixel]')
                   ylabel('Temperature [^oC]')
                   legend('m','n')
```

- slider designed for showing the next images in a sequence:

```
hObj(27)=uicontrol('Style', 'slider',...
                   'units','normalized',....
                   'FontUnits','normalized',...
                   'BackgroundColor',col,...
                   'Position', [0.01 0.01 0.98 0.05],...
                   'Min',1,'Max',480,'Value',1,...
                   'SliderStep',[1/100 1/100],...
                   'Parent',hObj(1),...
                   'Callback', 'THV_GUI2_sw(7)');
```

- button for automatic playing of the next images in a sequence:

```
hObj(28)=uicontrol('Style', 'pushbutton',...
                   'units','normalized',...
                   'FontUnits','normalized',...
                   'String', 'PLAY',...
                   'Position', [0.01 0.06 0.3 0.04],...
                   'Parent',hObj(1),...
                   'Callback', 'THV_GUI2_sw(9)',...
                   'BackgroundColor',col);
```

- buttons for manual setting of analysis of any selected image in a sequence:

```
hObj(29)=uicontrol('Style', 'pushbutton',...
                   'units','normalized',...
                   'FontUnits','normalized',...
                   'String', 'SET',...
                   'Position', [0.31 0.06 0.3 0.04],...
                   'Parent',hObj(1),...
                   'Callback', 'THV_GUI2_sw(8)',...
                   'BackgroundColor',col);
```

- buttons to stop automatic playing of consecutive images in a sequence:

```
hObj(30)=uicontrol('Style', 'pushbutton',...
                   'units','normalized',...
                   'FontUnits','normalized',...
                   'String', 'STOP',...
                   'Position', [0.61 0.06 0.37 0.04],...
                   'Parent',hObj(1),...
                   'Callback', 'THV_GUI2_sw(10)',...
                   'BackgroundColor',col);
```

- text box designed to display the current frame number, the mean values for the vertical and horizontal cross-section:

```
hObj(31)=uicontrol('Style', 'text',...
                   'units','normalized',...
                   'String', {'FrameNumber = -',
  'Median_x = -','Median_y = -'},...
                   'Position', [0.85 0.15 0.15 0.29],...
                   'BackgroundColor',col,...
                   'Parent',hObj(1));
```

- button to play automatically the analysis of the patient's back:

```
hObj(32)=uicontrol('Style', 'pushbutton',...
                   'units','normalized',...
                   'FontUnits','normalized',...
                   'String', 'Back',...
                   'Position', [0.31 0.95 0.15 0.05],...
                   'Parent',hObj(1),...
                   'Callback', 'THV_GUI2_sw(11)',...
                   'BackgroundColor',col);
```

- button to play automatically the analysis of the area subjected to laser treatment:

```
hObj(33)=uicontrol('Style', 'pushbutton',...
                   'units','normalized',...
                   'FontUnits','normalized',...
                   'String', 'Laser',...
                   'Position', [0.46 0.95 0.15 0.05],...
                   'Parent',hObj(1),...
                   'Callback', 'THV_GUI2_sw(12)',...
                   'BackgroundColor',col);
```

- button for automatic playing of segmentation:

```
hObj(34)=uicontrol('Style', 'pushbutton',...
                   'units','normalized',...
                   'FontUnits','normalized',...
                   'String', 'Segmentation',...
                   'Position', [0.61 0.95 0.15 0.05],...
                   'Callback', 'THV_GUI2_sw(13)',...
                   'Parent',hObj(1),...
                   'BackgroundColor',col);
```

- button to enable or disable automatic scaling on the graph of temperature changes:

```
hObj(35)=uicontrol('Style', 'radiobutton',...
                   'units','normalized',...
                   'FontUnits','normalized',...
                   'String', 'Axis auto',...
                   'Position', [0.61 0.95 0.15 0.05],...
                   'Callback', 'THV_GUI2_sw(14)',...
                   'Parent',hObj(23),...
                   'BackgroundColor',col);
```

- button to enable or disable indicating the ROI:

```
hObj(36)=uicontrol('Style', 'checkbox',...
                   'units','normalized',...
                   'FontUnits','normalized',...
                   'String', 'ROI on/off',...
                   'Position', [0.001 0.25 0.15 0.05],...
                   'Value',0,...
                   'Parent',hObj(1),...
                   'Callback','THV_GUI2_sw(15)',...
                   'BackgroundColor',col);
```

- creation of a new window containing the ROI image:

```
hObj(37)=figure('Name','ROI',...
                'NumberTitle','off');
hObj(39)=axes('Parent',hObj(37),...
              'units','normalized',...
              'Position', [0.1 0.1 0.85 0.85]);
hObj(38)=imshow(LGRAY,'InitialMagnification','fit',
'Parent',hObj(39));
                xlabel('n_{ROI} [pixel]')
                ylabel('m_{ROI} [pixel]')
                colormap('jet'); colorbar
```

- button to start and stop calculating GLCM in the ROI:

```
hObj(40)=uicontrol('Style', 'checkbox',...
                   'units','normalized',...
                   'FontUnits','normalized',...
                   'String', 'GLCM on/off',...
                   'Position', [0.001 0.3 0.15 0.05],...
                   'Value',0,...
                   'Parent',hObj(1),...
                   'Callback','THV_GUI2_sw(16)',...
                   'BackgroundColor',col);
```

- creation of a new window containing the GLCM calculated from the ROI image:

```
hObj(41)=figure('Name','ROI-GLCM',...
                'NumberTitle','off');
hObj(42)=axes('Parent',hObj(41),...
              'units','normalized',...
              'Position', [0.1 0.1 0.85 0.85]);
hObj(43)=imshow(LGRAY,'InitialMagnification','fit',
'Parent',hObj(42));
                   xlabel('gn_{GLCM} [pixel]')
                   ylabel('gm_{GLCM} [pixel]')
                   colormap('jet'); colorbar
```

- button to start and stop calculating square-tree decomposition in the ROI:

```
hObj(44)=uicontrol('Style', 'checkbox',...
                   'units','normalized',...
                   'FontUnits','normalized',...
                   'String', 'QTDCP on/off',...
                   'Position', [0.001 0.35 0.15 0.05],...
                   'Value',0,...
                   'Parent',hObj(1),...
                   'Callback','THV_GUI2_sw(17)',...
                   'BackgroundColor',col);
```

- creating a new window containing the square-tree decomposition matrix calculated from the ROI image:

```
hObj(45)=figure('Name','ROI-QTDCP',...
                'NumberTitle','off');
hObj(46)=axes('Parent',hObj(45),...
              'units','normalized',...
              'Position', [0.1 0.1 0.85 0.85]);
hObj(47)=imshow(LGRAY,'InitialMagnification','fit',
'Parent',hObj(46));
                   xlabel('qn_{GLCM} [pixel]')
                   ylabel('qm_{GLCM} [pixel]')
```

- text box for showing contrast, homogeneity, correlation and energy values calculated from the GLCM based on the ROI:

```
hObj(48)=uicontrol('Style', 'text',...
                   'units','normalized',...
                   'String', ['Contrast=- ',
'Homogeneity=- ','Correlation=- ','Energy=- '],...
                   'Position', [0.31 0.9 0.45 0.05],...
                   'BackgroundColor',col,...
                   'Parent',hObj(1));
```

The GUI created on this basis is shown in Fig. 6.17.

The presented m-file *THV_GUI_MAIN2* is closely related to the function *THV_GUI2_sw*. *THV_GUI2_sw* is an integral part and contains algorithms. They are available after pressing the button or selecting from the menu. The three buttons

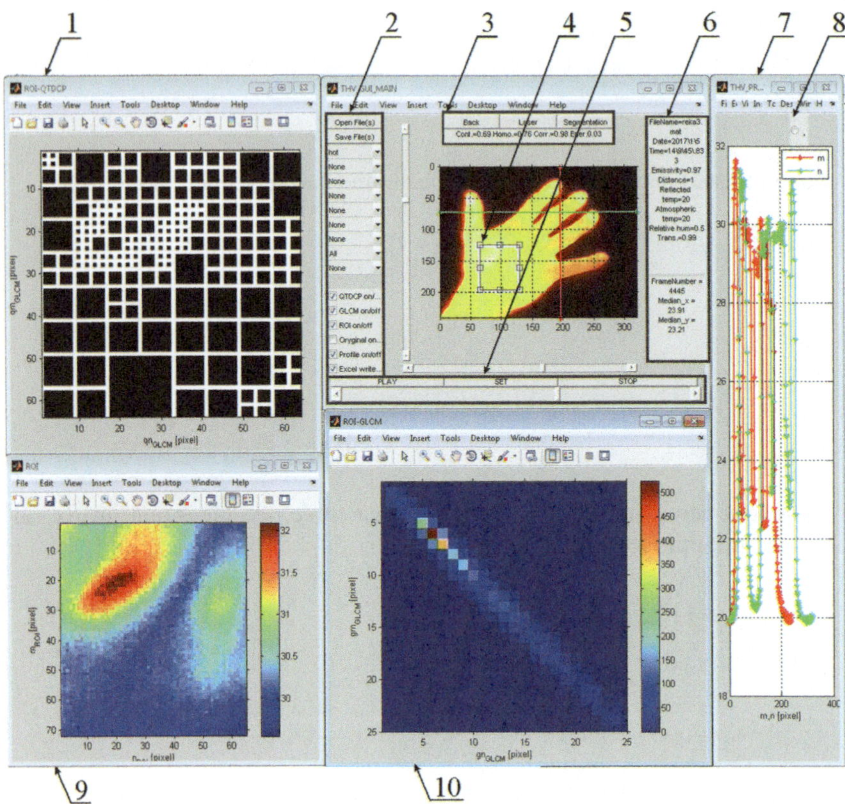

Fig. 6.17 GUI of the application designed for image sequence analysis: *1* window of square-tree decomposition results, *2* menu containing the known elements of processing and new on/off buttons for square-tree decomposition, analysis of GLCM and ROI; *3* buttons of the dedicated analysis of the back, laser effects and segmentation; *4* manually selected ROI; *5* menu items responsible for image sequences: playing, stopping and manual setting; *6* two text boxes—one showing the basic acquisition parameters and the other showing the results of current calculations; *7* window showing temperature changes for the horizontal and vertical cross-section; *8* button for automatic or manual setting of the range of values of each axis; *9* window showing the ROI; *10* window of the image of GLCM calculated from the selected ROI

associated with the described dedicated analysis of the patient's back, the area subjected to laser treatment and segmentation were not included intentionally in the following description of particular fragments of the function *THV_GUI2_sw*. This is left to the readers who certainly will be able to do it themselves on the basis of the source code shown above. Thus, the function *THV_GUI2_sw* contains the following elements:

- header elements:

```
function THV_GUI2_sw(sw)
%THV_sw graphical user interface
%
%   Copyright 2017 Robert Koprowski
%
%   Free only for readers book
```

- declaration of global variables:

```
global hObj PathName FileName LGRAY excel_data
Lvar bin_var nr_end nr_start L sonoff hObj_rect
```

- checking if *sw* = 1, if so, an attempt is made to read an image sequence:

```
if sw==1 % read mat file
    [FileName,PathName] = uigetfile('*.mat','Select MAT file');
    if FileName~=0
        if strcmp(FileName(end-2:end),'mat')
            L=load([PathName,FileName]);
            FieldName=fieldnames(L);
            NumberLastImageStr=FieldName{end-2};
            NumberFirstImageStr=FieldName{1};
```

- saving the initial value of the image number (does not have to be numbered from "1") and the number of the last image in the relevant variables:

```
nr_end=str2double(NumberLastImageStr(6:end));
        nr_start=str2double(NumberFirstImageStr(6:end));
```

- setting on the slider the value of changes of consecutive images and saving parameters such as time or emissivity in the variables:

```
set(hObj(27),'Max',nr_end,'Min',nr_start,'Value',nr_start);

        eval(['time=L.',FileName(1:end-4),'_DateTime;']);
        eval(['param=L.',FileName(1:end-4),'_ObjectParam;']);
        param=round(param*100)/100;
```

- saving the first image in the variable L_{GRAY}:

```
eval(['LGRAY=L.Frame',mat2str(nr_start),';']);
            set(hObj(19),'String', {['FileName=',FileName],...
```

- saving date and time as well as distance emissivity, reflectance and others in relevant variables:

```
['Date=',mat2str(time(1)),'\',mat2str(time(2)),'\',mat2str(time(3))],...
['Time=',mat2str(time(4)),'\',mat2str(time(5)),'\',mat2str(time(6)),'\.',
mat2str(time(7))],...
            ['Emissivity=',mat2str(param(1))],...
            ['Distance=',mat2str(param(2))],...
            ['Reflected temp=',mat2str(param(3))],...
            ['Atmospheric temp=',mat2str(param(4))],...
            ['Relative hum=',mat2str(param(5))],...
            ['Trans.=',mat2str(param(6))]})
        set(hObj(15),'Min',1,'Max',size(LGRAY,1));
        set(hObj(16),'Min',1,'Max',size(LGRAY,2));
        bin_var=(round( (min(LGRAY(:)):0.1:max(LGRAY(:)))*10 )./10)';
```

- reset and setting variables for default values:

```
set(hObj(6),'String',{'None';num2str(bin_var)})
        excel_data=[];
        Lvar=LGRAY;
        THV_GUI2_sw(4); % set slider
        THV_GUI2_sw(3);
    else
        disp('This version read only *.mat files')
    end
end
end
```

- checking if *sw* = 2, if so, an attempt is made to save a screenshot to a file:

```
if sw==2 % write file
    Lgetframe=getframe(gcf);
    clk=clock;
    imwrite(Lgetframe.cdata,[[PathName,FileName],date,
mat2str(clk(3:end)),'.jpg'])
    end
```

- checking if *sw* = 3, if so, there follows processing—selection of the colour palette:

```
if sw==3
    Lvar=LGRAY;
    if get(hObj(4),'Value')==1
        colormap('autumn')
    end
    if get(hObj(4),'Value')==2
        colormap('bone')
    end
    if get(hObj(4),'Value')==3
        colormap('colorcube')
    end
    if get(hObj(4),'Value')==4
        colormap('cool')
    end
    if get(hObj(4),'Value')==5
        colormap('copper')
    end
    if get(hObj(4),'Value')==6
        colormap('gray')
    end
    if get(hObj(4),'Value')==7
        colormap('hot')
    end
    if get(hObj(4),'Value')==8
        colormap('hsv')
    end
    if get(hObj(4),'Value')==9
        colormap('jet')
    end
    if get(hObj(4),'Value')==10
        colormap('lines')
    end
    if get(hObj(4),'Value')==11
        colormap('pink')
    end
    if get(hObj(4),'Value')==12
        colormap('prism')
    end
    if get(hObj(4),'Value')==13
        colormap('spring')
    end
    if get(hObj(4),'Value')==14
        colormap('summer')
    end
    if get(hObj(4),'Value')==15
        colormap('white')
    end
    if get(hObj(4),'Value')==16
        colormap('winter')
    end
```

- median filtering:

```
hs=get(hObj(5),'Value');
if hs>1
    Lvar=medfilt2(Lvar,[hs*2-1, hs*2-1],'symmetric');
end
```

- binarization:

```
pr=get(hObj(6),'Value');
if pr>1
    Lvar=Lvar>(bin_var(pr-1));
end
```

- dilation:

```
SE=get(hObj(7),'Value');
if SE>1
    Lvar=imdilate(Lvar,ones([SE*2-1, SE*2-1]));
end
```

- erosion:

```
SE=get(hObj(8),'Value');
if SE>1
    Lvar=imerode(Lvar,ones([SE*2-1, SE*2-1]));
end
```

- opening:

```
SE=get(hObj(9),'Value');
if SE>1
    Lvar=imopen(Lvar,ones([SE*2-1, SE*2-1]));
end
```

- closing

```
SE=get(hObj(10),'Value');
if SE>1
    Lvar=imclose(Lvar,ones([SE*2-1, SE*2-1]));
end
```

- selection of one, two or three largest objects:

```
if get(hObj(11),'Value')~=1
    if sum(sum(Lvar))>0
    if sum(sum((Lvar~=1)&(Lvar~=0)))==0
        Llab=bwlabel(Lvar); pam_pow=[];
        for il=1:max(Llab(:))
            pam_pow(il,1:2)=[sum(sum(Llab==il)) il];
        end
        pam_pow_l=sortrows(pam_pow,-1);
        if (size(pam_pow_l,1)>=1) && (get(hObj(11),'Value')==2)
            Lvar=Llab==pam_pow_l(1,2);
        end
        if (size(pam_pow_l,1)>=2) && (get(hObj(11),'Value')==3)
            Lvar=(Llab==pam_pow_l(1,2))|(Llab==pam_pow_l(2,2));
        end
        if (size(pam_pow_l,1)>=3) && (get(hObj(11),'Value')==4)
            Lvar=(Llab==pam_pow_l(1,2))|(Llab==pam_pow_l(2,2))|
(Llab==pam_pow_l(3,2));
        end
        if (size(pam_pow_l,1)>=1) && (get(hObj(11),'Value')==5)
            Lvar=Llab==pam_pow_l(end,2);
        end
    else
        warndlg('It''s NOT binary image')
    end
    end
end
```

- filling holes in the object:

```
if get(hObj(12),'Value')==2
    Lvar=bwfill(Lvar,'holes');
end
```

- filling holes in the object:

```
if  get(hObj(20),'Value')==0
    set(hObj(14),'CData',Lvar);
    if min(Lvar(:))~=max(Lvar(:))
        caxis(hObj(13),[min(Lvar(:)) max(Lvar(:))])
    end
    axis([0 size(Lvar,2) 0 size(Lvar,1)]);
else
    set(hObj(14),'CData',LGRAY);
    if min(LGRAY(:))~=max(LGRAY(:))
        caxis(hObj(13),[min(LGRAY(:)) max(LGRAY(:))])
    end
    axis([0 size(LGRAY,2) 0 size(LGRAY,1)]);
end
```

- updating data to be saved to Excel:

```
THV_GUI2_sw(4);
excel_data=double([min(Lvar(:)) max(Lvar(:))]);
THV_GUI2_sw(6);
```

- cutting the ROI based on the selected area:

```
if get(hObj(36),'Value')==1
    rect_=round(getPosition(hObj_rect));
    nrp=rect_(1);
    nrk=rect_(1)+rect_(3);
    mrp=rect_(2);
    mrk=rect_(2)+rect_(4);
    mrp(mrp<1)=1;
    nrp(nrp<1)=1;
    mrk(mrk>size(Lvar,1))=size(Lvar,1);
    nrk(nrk>size(Lvar,2))=size(Lvar,2);
    LvarROI=Lvar(mrp:mrk,nrp:nrk);
    set(hObj(38),'CData',LvarROI);
    axis(hObj(39),[1 size(LvarROI,2) 1 size(LvarROI,1)]);
    caxis(hObj(39),[min(LvarROI(:)) max(LvarROI(:))])
end
```

- calculating GLCM in the ROI:

```
if get(hObj(40),'Value')==1
    if exist('LvarROI')==1
        NumLevels=round((max(LvarROI(:))-min(LvarROI(:)))*10);
        LvarROIGLCM=graycomatrix(LvarROI,'GrayLimits',
[min(LvarROI(:)) max(LvarROI(:))],'NumLevels',NumLevels);
        set(hObj(43),'CData',LvarROIGLCM);
        axis(hObj(42),[1 size(LvarROIGLCM,2) 1 size(LvarROIGLCM,1)]);
        caxis(hObj(42),[min(LvarROIGLCM(:)) max(LvarROIGLCM(:))]);
```

- calculating contrast, homogeneity, energy and correlation based on the GLCM calculated for the ROI:

```
stats1_=graycoprops(LvarROIGLCM,{'contrast','homogeneity'});
        stats1(1)=stats1_.Contrast; stats1(2)=stats1_.Homogeneity;
        stats2_=graycoprops(LvarROIGLCM,{'correlation','energy'});
        stats2(1)=stats2_.Correlation; stats2(2)=stats2_.Energy;
        stats1=round(stats1*100)/100;
        stats2=round(stats2*100)/100;
        set(hObj(48),'String', ['Cont.=',mat2str(stats1(1)),'
Homo.=',mat2str(stats1(2)),' Corr.=',mat2str(stats2(1)),'
Ener.',mat2str(stats2(2))]);
    else
        warndlg('ROI not exist - please select ROI','! Warning !')
        set(hObj(48),'String', ['Contrast=- ','Homogeneity=-
','Correlation=- ','Energy=- ']);
    end
else
        set(hObj(48),'String', ['Contrast=- ',
'Homogeneity=- ','Correlation=- ','Energy=- ']);
end
```

- calculating square-tree decomposition in the ROI:

```
if get(hObj(44),'Value')==1
    if exist('LvarROI')==1
        LvarROI2=imresize(LvarROI,[2.^round(log2(size(LvarROI,1)))
2.^round(log2(size(LvarROI,2)))],'nearest');
        LvarROIQTDCMP=qtdecomp(LvarROI2,0.5);
        LvarROIQTDCMP2 = repmat(uint8(0),size(LvarROIQTDCMP));
        for dim = [512 256 128 64 32 16 8 4 2 1];
            numblocks=length(find(LvarROIQTDCMP==dim));
            if numblocks>0
                values=repmat(uint8(1),[dim dim numblocks]);
                values(2:dim,2:dim,:)=0;
                LvarROIQTDCMP2=qtsetblk(LvarROIQTDCMP2,
LvarROIQTDCMP,dim,values);
            end
        end
        set(hObj(47),'CData',LvarROIQTDCMP2);
        axis(hObj(46),[1 size(LvarROIQTDCMP2,2) 1
size(LvarROIQTDCMP2,1)]);
        caxis(hObj(46),[min(LvarROIQTDCMP2(:)) max(LvarROIQTDCMP2(:))])
    else
        warndlg('ROI not exist - please select ROI','! Warning !')
    end
end
end
```

- calculating the horizontal and vertical profile for different values of rows and columns indicated with the slider:

```matlab
   if sw==4
      ww=get(hObj(15),'Max')-round(get(hObj(15),'Value'));
      kk=round(get(hObj(16),'Value'));
      set(hObj(17),'XData',[kk kk],'YData',[1 size(Lvar,1)]);
      set(hObj(18),'YData',[ww ww],'XData',[1 size(Lvar,2)]);
      set(hObj(25),'YData',Lvar(:,kk),'XData',[1:size(Lvar,1)]);
      set(hObj(26),'YData',Lvar(ww,:),'XData',[1:size(Lvar,2)]);
      nr_act=round(get(hObj(27),'Value'));
      set(hObj(31),'String', {'FrameNumber =',mat2str(nr_act),
'Median_x =',mat2str(round(median(Lvar(:,kk))*100)/100),
'Median_y =',mat2str(round(median(Lvar(ww,:))*100)/100)});
   end
```

- enabling and disabling the display of profile analysis:

```matlab
if sw==5
   if  get(hObj(21),'Value')==1
       set(hObj([15:18,23]),'Visible','on');
   else
       set(hObj([15:18,23]),'Visible','off');
   end
end
```

- saving data to Excel:

```matlab
if sw==6
    if get(hObj(22),'Value')==1
    if ~isempty(excel_data)
       clk=clock;
       xlswrite([[PathName,FileName],date,'.xls'],excel_data);
    end
    end
end
```

- updating calculations and image for a new position of the slider:

```matlab
if sw==7
   nr_act=get(hObj(27),'Value');
   eval(['LGRAY=L.Frame',mat2str(round(nr_act)),';']);
   THV_GUI2_sw(3);
end
```

- manual setting of the current image number:

```matlab
if sw==8
   i_ = inputdlg({'Enter frame number'},'Input for frame number',
1,{ mat2str(round(get(hObj(27),'Value'))) });
   if ~isempty(i_)
      i__=str2num(i_{1});
      if isempty(i__)
         warndlg('Please enter the number - the decimal point',
'! Warning !')
      else
         if (i__<=nr_end) && (i__>0)
            nr_act=i__;
            set(hObj(27),'Value',nr_act);
            THV_GUI2_sw(3);
         else
            warndlg(['Please enter frame number >',mat2str(nr_start),
' and <',mat2str(nr_end)],'! Warning !')
         end
      end
   end
end
```

- playback and analysis of subsequent images after pressing the PLAY button:

```
if sw==9
    sonoff=1;
    if get(hObj(28),'Value')==1
        for nr_act=get(hObj(27),'Value'):nr_end
            if sonoff==1
                set(hObj(27),'Value',nr_act);
                THV_GUI2_sw(7);
                pause(0.03)
            else
                return
            end
        end
    else
        return
    end
end
```

- stopping the playback and analysis of successive images after pressing the STOP button:

```
if sw==10
    sonoff=0;
end
```

- automatic analysis of the patient`s back, laser effects and segmentation—fragment of the algorithm intended for readers—described in the previous chapters:

```
if sw==11 %Back
end
if sw==12 %Laser
end
if sw==13 %Segmentation
end
```

- enabling or disabling automatic selection of values for x and y axes on a graph showing the temperature values in the profiles:

```
if sw==14 % axis auto
    if get(hObj(35),'Value')==1
        axis(hObj(24),'auto')
    end
    if get(hObj(35),'Value')==0
        axis(hObj(24),'manual')
    end
end
```

- manual selection of the ROI:

```
if sw==15
    if get(hObj(36),'Value')==1
        hObj_rect = imrect(hObj(13));
    else
        delete(hObj_rect)
    end
end
```

- enabling and disabling the dispaly of GLCM analysis results:

```
if sw==16
    if get(hObj(40),'Value')==1
        set(hObj(41),'Visible','on')
    else
        set(hObj(41),'Visible','off')
    end
end
```

- enabling and disabling the display of square-tree decomposition results:

```
if sw==17
    if get(hObj(44),'Value')==1
        set(hObj(45),'Visible','on')
    else
        set(hObj(45),'Visible','off')
    end
end
```

The functionality of each menu item (shown in Fig. 6.17) has been discussed in detail in previous chapters.

Chapter 7
Algorithm Sensitivity to Parameter Changes

Assessment of the algorithm sensitivity to parameter changes is a very important element [193–197]. All algorithms, especially as complex as the ones presented here, have some dozens of parameters set manually or automatically. In the case of manual selection of the parameter values, the question arises how their change ranging from 5 to 10% can affect the results obtained. Of course, the best solution here is to make their selection completely automatic—but it is not possible in every case. Therefore, the analysis of the effect of parameter changes on the results obtained may be of paramount importance in the diagnosis. Another important element is training the staff involved in examinations. This training should include information and evidence related to the impact of settings (e.g. of an infrared camera) on the results so that if any of the parameters could not be met (e.g. perpendicular orientation of the camera relative to the subject), they would know beforehand how it affects the results. Because in each case, for almost every algorithm, assessment of the algorithm sensitivity to parameter changes puts it in a bad light, this issue is ignored [198, 199]. Therefore authors deliberately do not analyse the algorithm sensitivity to parameter changes. There is often no mention of this issue in the literature. Here, therefore, there is an example of the method for analysing the algorithm sensitivity to parameter changes.

For further discussion, let us assume that the following parameters are measured:

- T_{mean}—mean temperature,
- T_{min}—minimum temperature,
- T_{max}—maximum temperature,

And the following parameters are changed:

- Δn_{ROI}—horizontal position of the ROI,
- Δm_{ROI}—vertical position of the ROI,
- ΔM_{ROI}—vertical size of the ROI.

© Springer International Publishing AG 2018
R. Koprowski, *Processing Medical Thermal Images*, Studies in Computational Intelligence 716, DOI 10.1007/978-3-319-61340-6_7

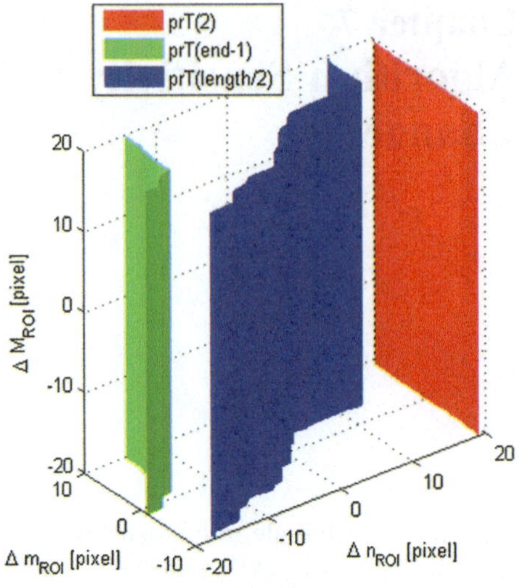

Fig. 7.1 Graph of changes in T_{min} for three thresholds when $\Delta n_{ROI} \in (-20,20)$, $\Delta m_{ROI} \in (-10,10)$, $\Delta M_{ROI} \in (-20,20)$

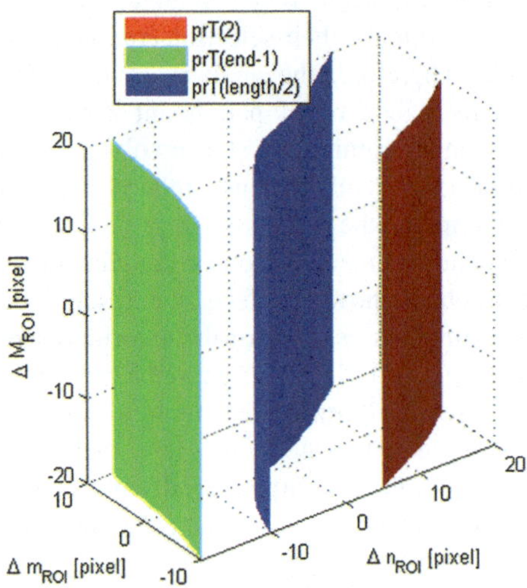

Fig. 7.2 Graph of changes in T_{mean} for three threshold when $\Delta n_{ROI} \in (-20,20)$, $\Delta m_{ROI} \in (-10,10)$, $\Delta M_{ROI} \in (-20,20)$

Figures 7.1, 7.2 and 7.3 show the obtained results in the form of planes plotted for three limits of p_{rT}. The values of p_{rT} are three selected values associated with the calculated T_{mean}, T_{min} and T_{max}. In addition, Fig. 7.4 shows a graph of changes in T_{mean} for the changes $\Delta n_{ROI} \in (-20,20)$, $\Delta m_{ROI} \in (-10,10)$ without changing the ROI size.

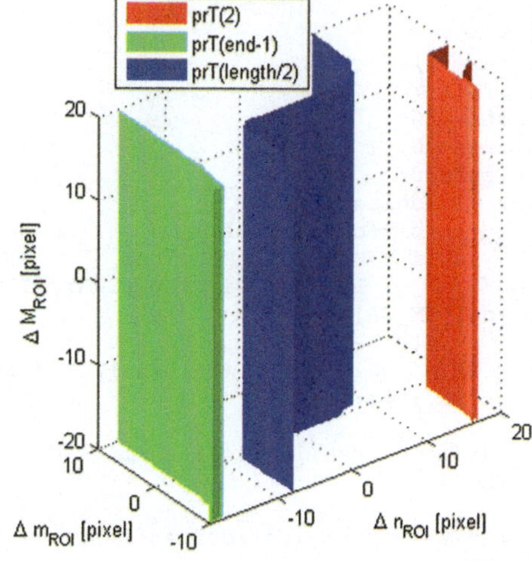

Fig. 7.3 Graph of changes in T_{max} for three thresholds when $\Delta n_{ROI} \in (-20,20)$, $\Delta m_{ROI} \in (-10,10)$, $\Delta M_{ROI} \in (-20,20)$

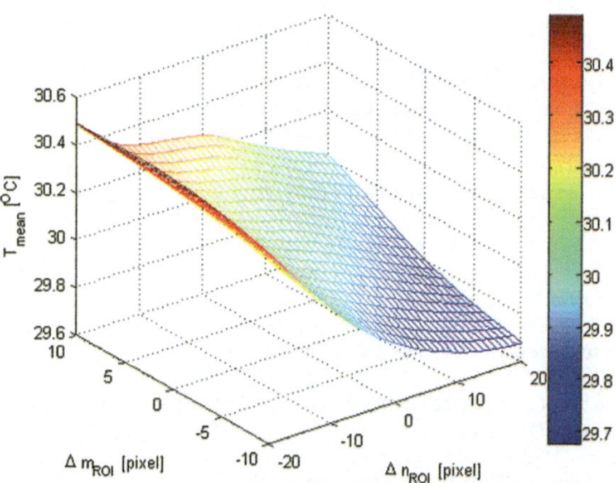

Fig. 7.4 Graph showing the relationship between changes in T_{mean} and changes in $\Delta n_{ROI} \in (-20,20)$, $\Delta m_{ROI} \in (-10,10)$ as well as $\Delta M_{ROI} = 0$

The above calculations were made using a simplified source code of the function *THV_GUI2_sw*, i.e.:

```
PathName='D:\';
FileName='reka3.mat';
L=load([PathName,FileName]);
FieldName=fieldnames(L);
NumberLastImageStr=FieldName{end-2};
NumberFirstImageStr=FieldName{1};
nr_end=str2double(NumberLastImageStr(6:end));
nr_start=str2double(NumberFirstImageStr(6:end));
eval(['LGRAY=L.Frame',mat2str(nr_start),';']);
figure; imshow(LGRAY,[]); colormap('jet')
LGRAYROI=imcrop(LGRAY,[116 121 15 20]);
figure; imshow(LGRAYROI,[]); colormap('jet')
xlabel('n_{ROI} [pixel]');
```

```matlab
ylabel('m_{ROI} [pixel]');
T_mean=[];
T_max=[];
T_min=[];
deltam=10;
deltan=20;
deltaM=20;
for mi=-deltam:deltam
    for ni=-deltan:deltan
        for Mi=-deltaM:deltaM
            LGRAYROI=imcrop(LGRAY,[116+ni 121+mi
15+deltaM 20]);

T_mean(mi+deltam+1,ni+deltan+1,Mi+deltaM+1)=mean(
LGRAYROI(:));

T_max(mi+deltam+1,ni+deltan+1,Mi+deltaM+1)=max(LG
RAYROI(:));

T_min(mi+deltam+1,ni+deltan+1,Mi+deltaM+1)=min(LG
RAYROI(:));
        end
    end
end
[xx,yy,zz]=meshgrid(-deltan:deltan,-
deltam:deltam,-deltaM:deltaM);
prT=min(T_min(:)):0.1:max(T_min(:));
figure
p1 = patch(isosurface(xx,yy,zz,T_min,prT(2)),...
    'FaceColor','red','EdgeColor','none');
isonormals(xx,yy,zz,T_min,p1)
p2 = patch(isosurface(xx,yy,zz,T_min,prT(end-
1)),...
    'FaceColor','green','EdgeColor','none');
isonormals(xx,yy,zz,T_min,p2)
p3 =
patch(isosurface(xx,yy,zz,T_min,prT(round(length(
prT)/2))),...
    'FaceColor','blue','EdgeColor','none');
isonormals(xx,yy,zz,T_min,p3)
view(3); daspect([1 1 1]); axis tight
camlight;  camlight(-80,-10); lighting phong;
grid on
xlabel('\Delta n_{ROI} [pixel]');
ylabel('\Delta m_{ROI} [pixel]');
zlabel('\Delta M_{ROI} [pixel]')
legend('prT(2)','prT(end-1)','prT(length/2)')
%%%%
prT=min(T_mean(:)):0.1:max(T_mean(:));
figure
p1 = patch(isosurface(xx,yy,zz,T_mean,prT(2)),...
    'FaceColor','red','EdgeColor','none');
isonormals(xx,yy,zz,T_mean,p1)
```

```
p2 = patch(isosurface(xx,yy,zz,T_mean,prT(end-
1))),...
    'FaceColor','green','EdgeColor','none');
isonormals(xx,yy,zz,T_mean,p2)
p3 =
patch(isosurface(xx,yy,zz,T_mean,prT(round(length
(prT)/2)))),...
    'FaceColor','blue','EdgeColor','none');
isonormals(xx,yy,zz,T_mean,p3)
view(3); daspect([1 1 1]); axis tight
camlight;  camlight(-80,-10); lighting phong;
grid on
xlabel('\Delta n_{ROI} [pixel]');
ylabel('\Delta m_{ROI} [pixel]');
zlabel('\Delta M_{ROI} [pixel]')
legend('prT(2)','prT(end-1)','prT(length/2)')
%%%%
prT=min(T_max(:)):0.1:max(T_max(:));
figure
p1 = patch(isosurface(xx,yy,zz,T_max,prT(2)),...
    'FaceColor','red','EdgeColor','none');
isonormals(xx,yy,zz,T_max,p1)
p2 = patch(isosurface(xx,yy,zz,T_max,prT(end-
1))),...
    'FaceColor','green','EdgeColor','none');
isonormals(xx,yy,zz,T_max,p2)
p3 =
patch(isosurface(xx,yy,zz,T_max,prT(round(length(
prT)/2)))),...
    'FaceColor','blue','EdgeColor','none');
isonormals(xx,yy,zz,T_max,p3)
view(3); daspect([1 1 1]); axis tight
camlight;  camlight(-80,-10); lighting phong;
grid on
xlabel('\Delta n_{ROI} [pixel]');
ylabel('\Delta m_{ROI} [pixel]');
zlabel('\Delta M_{ROI} [pixel]')
legend('prT(2)','prT(end-1)','prT(length/2)')
```

The following range of variation of the parameters $\Delta n_{ROI} \in (-20,20)$, $\Delta m_{ROI} \in (-10,10)$, $\Delta M_{ROI} \in (-20,20)$ was adopted. It can be freely changed. For the given range of variation, calculations for one of the images shown in Figs. 7.1, 7.2 or 7.3 take about 2.8 s (for the Intel® Core i7 3.6 GHz, 64 GB RAM). Interpretation of the results obtained is intuitive. For example, the results shown in Fig. 7.4 are used to determine ranges of variation of T_{mean}. This temperature varies in the range from 29.7 to 30.5 °C when the position of the ROI is changed in the range of $\Delta n_{ROI} \in (-20,20)$, $\Delta m_{ROI} \in (-10,10)$. In addition, the presented graph (Fig. 7.4) shows that a change in both Δn_{ROI} and Δm_{ROI} affects the results in a similar way. Therefore, it can be said that the direction of the ROI shift in rows or columns does not matter—and so the mean temperature may range from 29.7 to 30.5 °C. Formally, the relative value of sensitivity $TPR\Delta_{Tmean}$ can be assessed with the following formula:

$$TPR_{T_{mean}} = \frac{\left| T_{mean}\left(\Delta n_{ROI}^{(2)}\right) - T_{mean}\left(\Delta n_{ROI}^{(1)}\right) \right|}{\left| \Delta n_{ROI}^{(2)} - \Delta n_{ROI}^{(1)} \right|} \cdot \frac{\Delta n_{ROI}^{(1)}}{T_{mean}\left(\Delta n_{ROI}^{(1)}\right)} \qquad (7.1)$$

where:

$\Delta n_{ROI}^{(1)}$ base shift, most frequently equal to zero,
$\Delta n_{ROI}^{(2)}$ specified shift,
$T_{mean}(\Delta n_{ROI}^{(1)})$ mean temperature for the base shift—most frequently equal to zero,
$T_{mean}(\Delta n_{ROI}^{(2)})$ mean temperature for the specified shift

However, in practice, absolute values of changes in the size measured depending on the parameter(s) change are provided.

The second example is the sensitivity assessment of the measurement of contrast, energy, correlations calculated form the GLCM. Similarly to the previous example, the results are shown in Figs. 7.5, 7.6, 7.7 and 7.8.

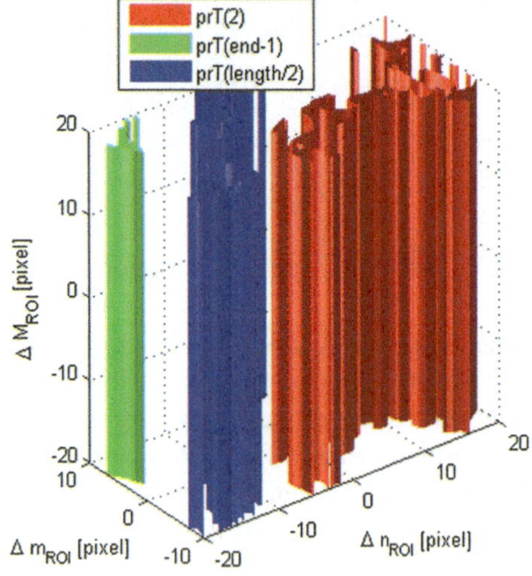

Fig. 7.5 Graph of contrast changes for three thresholds when $\Delta n_{ROI} \in (-20,20)$, $\Delta m_{ROI} \in (-10,10)$, $\Delta M_{ROI} \in (-20,20)$

Fig. 7.6 Graph of homogeneity changes for one threshold when $\Delta n_{ROI} \in (-20,20)$, $\Delta m_{ROI} \in (-10,10)$, $\Delta M_{ROI} \in (-20,20)$

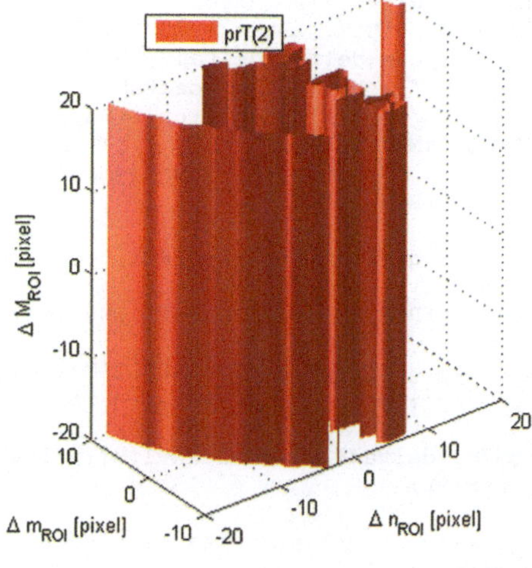

Fig. 7.7 Graph of correlation changes for one threshold when $\Delta n_{ROI} \in (-20,20)$, $\Delta m_{ROI} \in (-10,10)$, $\Delta M_{ROI} \in (-20,20)$

Fig. 7.8 Graph showing the relationship between correlation changes and changes in $\Delta n_{ROI} \in (-20,20)$, $\Delta m_{ROI} \in (-10,10)$ and $\Delta M_{ROI} = 0$

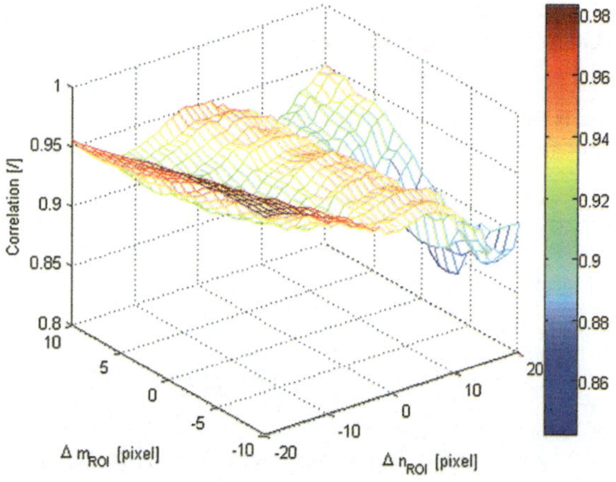

A dedicated source code (similar to the previous one) is as follows:

```matlab
tic
PathName='D:\';
FileName='reka3.mat';
L=load([PathName,FileName]);
FieldName=fieldnames(L);
NumberLastImageStr=FieldName{end-2};
NumberFirstImageStr=FieldName{1};
nr_end=str2double(NumberLastImageStr(6:end));
nr_start=str2double(NumberFirstImageStr(6:end));
eval(['LGRAY=L.Frame',mat2str(nr_start),';']);
figure; imshow(LGRAY,[]); colormap('jet')
LGRAYROI=imcrop(LGRAY,[116 121 15 20]);
figure; imshow(LGRAYROI,[]); colormap('jet')
xlabel('n_{ROI} [pixel]');
ylabel('m_{ROI} [pixel]');
con=[];
hom=[];
cor=[];
ene=[];
deltam=10;
deltan=20;
deltaM=20;
for mi=-deltam:deltam
    for ni=-deltan:deltan
        for Mi=-deltaM:deltaM
            LvarROI=imcrop(LGRAY,[116+ni 121+mi
15+deltaM 20]);
            NumLevels=round((max(LvarROI(:))-
min(LvarROI(:)))*10);
            Lvar-
ROIGLCM=graycomatrix(LvarROI,'GrayLimits',[min(Lv
arROI(:))
max(LvarROI(:))],'NumLevels',NumLevels);

stats1_=graycoprops(LvarROIGLCM,{'contrast','homo
geneity'});

con(mi+deltam+1,ni+deltan+1,Mi+deltaM+1)=stats1_.
Contrast;

hom(mi+deltam+1,ni+deltan+1,Mi+deltaM+1)=stats1_.
Homogeneity;

stats2_=graycoprops(LvarROIGLCM,{'correlation','e
nergy'});

cor(mi+deltam+1,ni+deltan+1,Mi+deltaM+1)=stats2_.
Correlation;

ene(mi+deltam+1,ni+deltan+1,Mi+deltaM+1)=stats2_.
Energy;
        end
    end
end
[xx,yy,zz]=meshgrid(-deltan:deltan,-
deltam:deltam,-deltaM:deltaM);
prT=min(con(:)):0.1:max(con(:));
toc
figure
p1 = patch(isosurface(xx,yy,zz,con,prT(2)),...
```

```matlab
    'FaceColor','red','EdgeColor','none');
isonormals(xx,yy,zz,con,p1)
p2 = patch(isosurface(xx,yy,zz,con,prT(end-
1)),...
    'FaceColor','green','EdgeColor','none');
isonormals(xx,yy,zz,con,p2)
p3 =
patch(isosurface(xx,yy,zz,con,prT(round(length(pr
T)/2))),...
    'FaceColor','blue','EdgeColor','none');
isonormals(xx,yy,zz,con,p3)
view(3); daspect([1 1 1]); axis tight
camlight;  camlight(-80,-10); lighting phong;
grid on
xlabel('\Delta n_{ROI} [pixel]');
ylabel('\Delta m_{ROI} [pixel]');
zlabel('\Delta M_{ROI} [pixel]')
legend('prT(2)','prT(end-1)','prT(length/2)')
%%%%
prT=min(hom(:)):0.1:max(hom(:));
figure
p1 = patch(isosurface(xx,yy,zz,hom,prT(2)),...
    'FaceColor','red','EdgeColor','none');
isonormals(xx,yy,zz,hom,p1)
p2 = patch(isosurface(xx,yy,zz,hom,prT(end-
1)),...
    'FaceColor','green','EdgeColor','none');
isonormals(xx,yy,zz,hom,p2)
p3 =
patch(isosurface(xx,yy,zz,hom,prT(round(length(pr
T)/2))),...
    'FaceColor','blue','EdgeColor','none');
isonormals(xx,yy,zz,hom,p3)
view(3); daspect([1 1 1]); axis tight
camlight;  camlight(-80,-10); lighting phong;
grid on
xlabel('\Delta n_{ROI} [pixel]');
ylabel('\Delta m_{ROI} [pixel]');
zlabel('\Delta M_{ROI} [pixel]')
legend('prT(2)','prT(end-1)','prT(length/2)')
%%%%
prT=min(cor(:)):0.1:max(cor(:));
figure
p1 = patch(isosurface(xx,yy,zz,cor,prT(2)),...
    'FaceColor','red','EdgeColor','none');
isonormals(xx,yy,zz,cor,p1)
p2 = patch(isosurface(xx,yy,zz,cor,prT(end-
1)),...
    'FaceColor','green','EdgeColor','none');
isonormals(xx,yy,zz,T_max,p2)
p3 =
patch(isosurface(xx,yy,zz,cor,prT(round(length(pr
T)/2))),...
    'FaceColor','blue','EdgeColor','none');
isonormals(xx,yy,zz,cor,p3)
view(3); daspect([1 1 1]); axis tight
```

```
camlight;   camlight(-80,-10); lighting phong;
grid on
xlabel('\Delta n_{ROI} [pixel]');
ylabel('\Delta m_{ROI} [pixel]');
zlabel('\Delta M_{ROI} [pixel]')
legend('prT(2)','prT(end-1)','prT(length/2)')

figure; mesh(xx(:,:,10),yy(:,:,10),cor(:,:,10));
colormap('jet'); colorbar
xlabel('\Delta n_{ROI} [pixel]');
ylabel('\Delta m_{ROI} [pixel]');
zlabel('Correlation [/]')
```

Time analysis for the images in Figs. 7.5, 7.6, 7.7 and 7.8 for the Intel® Core i7
3.6 GHz, 64 GB RAM is 34 s. At this point I encourage readers to test sensitivity
of the presented algorithm to other parameter changes.

In each case, sensitivity analysis should include potential negligence and
non-observance of procedures by the operator. These are parameters such as:
ensuring perpendicular orientation of the camera relative to the imaged area of the
skin, camera settings and even patient's preparation. It will be then possible, for
example, to answer the question how shorter time of patient's thermal adaptation
affects the results obtained.

Chapter 8
Conclusions

Analysis and processing of infrared images requires dedicated methods and algorithms. At present, despite the development of information technology [200]–[203], there is no universal method for the analysis of any images of any content. The algorithms that allow for complete measurement automation must be each time tailored to every new task. The advantage of tailoring algorithms is the above-mentioned full automation of their operation. Tens or hundreds of images can be analysed, for example, in batch mode. Unfortunately, the process of tailoring algorithms, and more specifically preparation of a new algorithm for each research task, is not an easy task. There is only a limited range of indications which elements such an algorithm can contain—e.g. median filtering in image pre-processing. Most of the remaining steps of the algorithm is usually designed ad hoc, taking into account the medical evidence and the designer's experience. Unfortunately, despite the universality of image analysis and processing in everyday life, there is still no universal approach allowing for fully automated image analysis. Admittedly in the area of thermal images some of the transformations which are difficult are simplified, e.g. in the analysis of images in visible light. However, there are several points in the analysis causing difficulties. This applies to segmentation that can be carried out by means of ordinary binarization—if the restrictions on image acquisition are respected. The separation of individual areas of the body is also extremely difficult, e.g. when the hand is touching the torso, it is not possible to separate them (due to temperature equalisation). In addition, there are problems typical of image processing related to measurement automation that must be preserved for a large variability of individual patients. Therefore, once again I encourage readers to test the presented algorithms on different types of infrared images and test their sensitivity to changing parameters. It should be also clearly stressed here that the author of this monograph does not assume any responsibility for incorrect and erroneous operation of the described algorithms.

© Springer International Publishing AG 2018

R. Koprowski, *Processing Medical Thermal Images*, Studies in Computational Intelligence 716, DOI 10.1007/978-3-319-61340-6_8

Appendix

The m-files and Matlab functions presented in this monograph are available at http://extras.springer.com/. As a result, an interested reader will not have to rewrite the source code from the book. The m-files are listed below in the order they are discussed in the text, i.e.:

- the first two concern the basic version of the GUI together with the necessary algorithms for the analysis of a single image:

THV_GUI_MAIN   m-file associated with GUI—image analysis
THV_GUI_sw   function containing algorithms for image analysis

- auxiliary m-files and other algorithms discussed in the book:

back1   algorithms related to the analysis of the patient's back,
laser1   algorithm allowing for testing segmentation methods of images associated with laser effects on the human skin based on an artificial image,
laser2   analysis of the sample image of the skin treated with laser,
segmentation1   algorithms associated with the proposed segmentation method,
stab1   algorithms for image stabilization,
dergauss   Gaussian function and its derivatives,
serpentine1   algorithms for serpentine analysis—part 1,
serpentine2   algorithms for serpentine analysis—part 2,
ident1   algorithms for object identification,
sens1   algorithm for sensitivity measurement—part 1
sens2   algorithm for sensitivity measurement—part 2.

- the last two refer to the advanced version of the GUI together with the necessary algorithms for image sequence analysis:

THV_GUI_MAIN2   m-file associated with the GUI—image sequence analysis
THV_GUI2_sw   function containing algorithms for image sequence analysis

© Springer International Publishing AG 2018
R. Koprowski, *Processing Medical Thermal Images*, Studies in Computational
Intelligence 716, DOI 10.1007/978-3-319-61340-6

References

1. Barnes R, Gershon-Cohen J. Clinical thermography. Jama J Am Med Assoc. 1963;185 (12):949–52.
2. Webster J. (Ed.) Medical instrumentation: application and design. Boston: Houghton Mifflin; 1978.
3. Vreugdenburg Thomas D, et al. A systematic review of elastography, electrical impedance scanning, and digital infrared thermography for breast cancer screening and diagnosis. Breast cancer Res Treat. 2013;137(3):665–76.
4. Moskowitz M, Feig SA, Cole-Beuglet C, Fox SH, Haberman JD, Libshitz HI, Zermeno A. Evaluation of new imaging procedures for breast cancer; proper process. Am J Roentgenol. 1983;140:591–94.
5. Francis J. Ring, Pioneering progress in infrared imaging in medicine. Quant InfraRed Thermography. 2014.
6. Clasen S, Pereira P, Lubienski A et al. CT- and MR-guided interventions in radiology interventional oncology, p. 205–62, Jan 1, 2013.
7. Jolesz F: History of image-guided therapy at brigham and women's hospital. Intraoperative Imaging and Image-Guided Therapy; 2014. p. 25–5.
8. Menzel R. Back matter—photonics, photonics, 2001.
9. Kaiser WA. Back matter—signs in MR-mammography. Signs in MR-Mammography, 2008.
10. Murray A, Pauling JD. Non-invasive methods of assessing Raynaud's Phenomenon. Raynaud's Phenomenon; 2015.
11. Lee K, Dretzke J, Grover L, Logan A, Moiemen N. A systematic review of objective burn scar measurements. Burns Trauma. 2016;4:1–33.
12. Ni Y, Mulier S, Miao Y, Michel L, Marchal G. A review of the general aspects of radiofrequency ablation. Abdom Imaging. 2005;30:381–400.
13. Wang LW, Peng CW, Chen C, Li Y. Quantum dots-based tissue and in vivo imaging in breast cancer researches: current status and future perspectives. Breast Cancer Res Treat. 2015;151:7–17.
14. Crocetti L, Della P, Clotilde C, et al. Peri-intraprocedural imaging: US, CT, and MRI. Abdom Imaging. 2011;36:648–60.
15. Filippo M, Bozzetti F, Martora R, Zagaria R, et al. Radiofrequency thermal ablation of renal tumors. La radiologia medica. 2014;119:499–511.
16. Boutin R, Powell, Sean T, Bracker M. Muscle, magnetic resonance imaging in orthopedic sports medicine; 2008. p. 1–44.
17. Ordon M, Findeiss L, Landman J. MR-guided renal ablation. Adv Image-Guided Urol Surg. 2015; 185–99.
18. Memarian N, Venetsanopoulos A, Chau T. Infrared thermography as an access pathway for individuals with severe motor impairments. J NeuroEng Rehabil. 2009;6:1–8.

19. Temple M, Waspe A, Amaral J, Napoli A, LeBlang S, Ghanouni P, et al. Establishing a clinical service for the treatment of osteoid osteoma using magnetic resonance-guided focused ultrasound: overview and guidelines. J Ther Ultrasound. 2016;4:1–11.

20. Gombos EC, Kacher DF, Caragacianu DL, Jayender J, Golshan M. Magnetic resonance imaging-guided breast intervention and surgery, intraoperative imaging and image-guided therapy; 2014. p. 817–44, 1 Jan 2014.

21. Borsook D, Becerra L. How close are we in utilizing functional neuroimaging in routine clinical diagnosis of neuropathic pain? Current Pain Headache Rep. 15:223–29, 1 June 2011.

22. Nayak KS, Nielsen J-F, Bernstein, MA, Markl M, Gatehouse P, Botnar R, Saloner D, Lorenz C, Wen H, Hu B, Epstein FH, Oshinski J, Raman SV. Cardiovascular magnetic resonance phase contrast imaging. J Cardiovasc Magn Reson 17:1–26, 9 Aug 2015.

23. Woodrum DA, Kawashima A, Gorny KR, Mynderse LA. Targeted prostate biopsy and MR-guided therapy for prostate cancer. Abdom Radiol. 41:877–88, 1 May 2016.

24. Tang S, Du Q, Liu T, Tan L, Niu M, Gao L, Huang Z, Fu C, Ma T, Meng X, Shao H. In vivo magnetic resonance imaging and microwave thermotherapy of cancer using novel Chitosan Microcapsules. Nanoscale Res Lett. 11:1–8, 5 July 2016.

25. Zhang S, Wang K, Liu H, Leng C, Gao Y, Tian J. Reconstruction method for in vivo bioluminescence tomography based on the split Bregman iterative and surrogate functions. Mol Imaging Biol. 1–11, 31 Aug 2016.

26. Xue J, Wu W, Liang P. Three-dimensional visualization technology and therapy planning system for microwave ablation therapy of liver tumor. In: Microwave ablation treat solid tumors; p. 283–92, 1 Jan 2015.

27. Qu P, Yu, Xiao-ling; Liang, Ping; Cheng, Zhigang; Han, Zhiyu; Liu, Fangyi; Yu, Jie, Application of contrast-enhanced ultrasound in the evaluation of clinical effect of microwave ablation of Hepatocellular Carcinoma: comparison with other imaging modalities. In: Microwave ablation treatment of solid tumors; p. 321–29, 1 Jan 2015.

28. Clasen S, Pereira PL, Lubienski A, Schäfer A-O, Mahnken AH, Helmberger T, Mack MG, Eichler K, Vogl TJ, Rosenberg C, Schiffman SC, Martin RCG, Baère T, Bruners P, Düx M, Mohnike K, Ricke J, Ditter P, Wilhelm KE, Strunk H, Beck A, Hengst S, Erinjeri JP, Gast T. Interventional oncology, CT- and MR-guided interventions in radiology; p. 205–62, 1 Jan 2013.

29. Bazrafshan B, Hübner F, Farshid P, Hammerstingl R, Paul J, Vogel V, Mäntele W, Vogl TJ. Temperature imaging of laser-induced thermotherapy (LITT) by MRI: evaluation of different sequences in phantom. Lasers Med Sci. 29:173–83, 1 Jan 2014.

30. Tsoumakidou G, Koch G, Caudrelier J, Garnon J, Cazzato RL, Edalat F, Gangi A. Image-guided spinal ablation: a review. Cardiovasc Intervent Radiol. 39:1229–238, 1 Sept 2016.

31. Helmberger T, Martí-Bonmatí L, Pereira P, Gillams A, Martínez J, Lammer J, Malagari K, Gangi A, Baere T, Adam EJ, Rasch C, Budach V, Reekers JA. Radiologists' leading position in image-guided therapy. Insights Imaging. 4:1–7, 1 Feb 2013.

32. Crocetti L, Della Pina C, Cioni D, Lencioni R. Peri-intraprocedural imaging: US, CT, and MRI. Abdom Imaging. 36:648–60, 1 Dec 2011.

33. Hong CW, Chow L, Turkbey EB, Lencioni R, Libutti SK, Wood BJ. Imaging features of radiofrequency ablation with heat-deployed liposomal doxorubicin in hepatic tumors. Cardiovasc Intervent Radiol. 39:409–16, 1 Mar 2016.

34. Wendler JJ, Pech M, Blaschke S, Porsch M, Janitzky A, Ulrich M, Dudeck O, Ricke J, Liehr U-B. Angiography in the isolated perfused kidney: radiological evaluation of vascular protection in tissue ablation by nonthermal irreversible electroporation. Cardiovasc Intervent Radiol. 35:383–90, 1 Apr 2012.

35. Denecke, T, Rühl R, Hildebrandt B, Stelter L, Grieser C, Stiepani H, Werk M, Podrabsky P, Plotkin M, Amthauer H, Ricke J, Lopez Hänninen E. Planning transarterial radioembolization of colorectal liver metastases with Yttrium 90 microspheres: evaluation of a sequential diagnostic approach using radiologic and nuclear medicine imaging techniques. Eur Radiol. 18:892–02, 1 May 2008.

36. Howe JY, Allard LF, Bigelow WC, Demers H, Overbury SH. Understanding catalyst behavior during in situ heating through simultaneous secondary and transmitted electron imaging. Nanoscale Res Lett. 9:1–6, 14 Nov 2014.

37. Cai W, Hsu AR, Li Z-B, Chen X. Are quantum dots ready for in vivo imaging in human subjects? Nanoscale Res Lett. 2:265–81, 30 May 2007.

38. Hahn MA, Singh AK, Sharma P, Brown SC, Moudgil BM. Nanoparticles as contrast agents for in-vivo bioimaging: current status and future perspectives. Anal Bioanal Chem. 399:3–27, 1 Jan 2011.

39. Würth C, Geißler D, Behnke T, Kaiser M, Resch-Genger U. Critical review of the determination of photoluminescence quantum yields of luminescent reporters. Anal Bioanal Chem. 407:59–8, 1 Jan 2015.

40. Leung A, Shukla S, Li E, Duann J-R, Yaksh T. Supraspinal characterization of the thermal grill illusion with fMRI. Mol Pain. 10:1–15, 11 Mar 2014.

41. Borsook D, Upadhyay J, Chudler EH, Becerra L. A key role of the basal ganglia in pain and analgesia—insights gained through human functional imaging. Mol Pain. 6:1–17, 13 May 2010.

42. Bazrafshan B, Hübner F, Farshid P, Hammerstingl R, Paul J, Vogel V, Mäntele W, Vogl TJ. Temperature imaging of laser-induced thermotherapy (LITT) by MRI: evaluation of different sequences in phantom. Lasers Med Sci. 29:173–83, 1 Jan 2014.

43. Jiang SC, Zhang XX. Dynamic modeling of photothermal interactions for laser-induced interstitial thermotherapy: parameter sensitivity analysis. Lasers Med Sci. 20:122–31, 1 Dec 2005.

44. Permpongkosol S, Link RE, Kavoussi LR, Solomon SB. Temperature measurements of the low-attenuation radiographic ice ball during CT-guided renal cryoablation, Cardiovasc Intervent Radiol. 31:116–21, 1 Jan 2008.

45. Power Sarah, Slattery MM, Lee MJ. Nanotechnology and its relationship to interventional radiology. Part II: drug delivery, thermotherapy, and vascular intervention. Cardiovasc Intervent Radiol. 34:676–90, 1 Aug 2011.

46. Maldague, X; Marinetti, S. Pulse phase infrared thermography. J Appl Phys. 79(5):2694–698 Published: 1 Mar 1996.

47. Jones, BF. A reappraisal of the use of infrared thermal image analysis in medicine, IEEE Trans Med Imaging. 17(6):1019–027 Published: Dec 1998.

48. U.S. Food And Drug Administration. Safety communication: breast cancer screening—thermography is not an alternative to mammography. Accessed 30 April 2013. Available from: URL: http://www.fda.gov/MedicalDevices/Safety/AlertsandNotices/ucm257259.htm.

49. Croatian Society of Radiology. CSR board public statement about thermography. Accessed 30 April 2013. Available from: URL www.radiologija.org/attachments/article/75/Priopcenje-HDR.pdf.

50. Croatian Society of Senology. CSS thermography statement. Accessed 30 April 2013. Available from: URL.

51. Hildebrandt C, Raschner C, Ammer K. An overview of recent application of medical infrared thermography in sports medicine in Austria. Sens (Basel). 2010;10(5):4700–15.

52. Ammer K. Diagnosis of Raynaud's phenomenon by thermography. Skin Res Technol. 1996;2(4):182–5.

53. Ammer K. The Glamorgan protocol for recording and evaluation of thermal images of the human body. Thermol Int. 2008;18:125–9.

54. Maldague XPV, Jones TS, Kaplan H, Marinetti S, Prystay M. Nondestructive handbook, infrared and thermal testing. ASNT Press; Columbus, OH, USA: 2001. Fundamentals of Infrared and Thermal Testing; p. 718.

55. Ring EFJ, Ammer K. The technique of infrared imaging in medicine. Thermol Int. 2000;10:7–14.

56. Plassmann P, Murawski P. CTHERM for standardized thermography. Proceedings of Abstracts, the 9th Congress of Thermology Poland; Krakow, Poland. May 29–June 1, 2003; p. 27–9.

57. Ring EFJ, Ammer K. Thermal Imaging in sports medicine. Sport Med Today. 1998;1:108–9.

58. Merla A, Mattei PA, Di Donato L, Romani GL. Thermal Imaging of cutaneous temperature modifications in runners during graded exercise. Ann Biomed Eng. 2010;38:158–63.

59. Schaefer G, Tait R, Zhu SY. Overlay of thermal and visual image using skin detection and image registration. Eng Med Biol Soc (EMBS). 2006;3:965–7.

60. Will RK, Ring EF, Clarke AK, Maddison PJ. Infrared thermography: what is its place in rheumatology in the 1990s? Br J Rheumatol. 1992 May;31(5):337–44. Review. No abstract available.

61. Will RK, Ring EF. The use of infra-red thermography in a rheumatology unit. Br J Rheumatol. 1991;30(1):71–2.

62. Reilly PA, Clarke AK, Ring EF. Thermography in carpal tunnel syndrome (CTS). Br J Rheumatol. 1989;28(6):553–4.

63. Bacon PA, Davies J, Ring FJ. Benoxaprofen–dose-range studies using quantitative thermography. J Rheumatol Suppl. 1980;6:48–53.

64. Bird HA, Ring EF, Bacon PA. Lack of correlation of arthroscopic synovial appearance with thermography in joint inflammation. Rheumatol Rehabil. 1978;17(2):107–12.

65. Kaczmarek M, Nowakowski A, Siebert J, Rogowski J. Infrared thermography: applications in heart surgery. Proc SPIE. 1999;3730:184–8.

66. Bird HA, Ring EF. Thermography and radiology in the localization of infection. Rheumatol Rehabil. 1978;17(2):103–6.

67. Robichaud G1, Garrard KP, Barry JA, Muddiman DC. MSiReader: an open-source interface to view and analyze high resolving power MS imaging files on Matlab platform. J Am Soc Mass Spectrom. 2013 May;24(5):718-21. doi: 10.1007/s13361-013-0607-z. Epub 2013 Mar 28.

68. Koprowski R. Blood pulsation measurement using cameras operating in visible light: limitations. Biomed Eng Online. 2016;15(1):111.

69. Koprowski R, Lanza M, Irregolare C. Corneal power evaluation after myopic corneal refractive surgery using artificial neural networks. Biomed Eng Online. 2016;15(1):121.

70. Koprowski R, Olczyk P. Segmentation in dermatological hyperspectral images: dedicated methods. Biomed Eng Online. 2016;15(1):97.

71. Konglerd P, Reeb C, Jansson F, Kaandorp JA. Quantitative morphological analysis of 2D images of complex-shaped branching biological growth forms: the example of branching thalli of liverworts. BMC Res Notes. 2017;10(1):103.

72. Peters JF, Ramanna S, Tozzi A, İnan E. Bold-independent computational entropy assesses functional donut-like structures in brain fMRI images. Front Hum Neurosci. 2017;1(11):38.

73. Schmidt FP, Hofer F, Krenn JR. Spectrum image analysis tool—a flexible MATLAB solution to analyze EEL and CL spectrum images. Micron. 2017;93:43–51.

74. Parry RM, Galhena AS, Gamage CM, Bennett RV, Wang MD, Fernández FM. omniSpect: an open MATLAB-based tool for visualization and analysis of matrix-assisted laser desorption/ionization and desorption electrospray ionization mass spectrometry images. J Am Soc Mass Spectrom. 2013;24(4):646–9.

75. Ferrer-Buedo J, Martínez-Sober M, Alakhdar-Mohmara Y, Soria-Olivas E, Benítez-Martínez JC, Martínez-Martínez JM. Matlab-based interface for the simultaneous acquisition of force measures and Doppler ultrasound muscular images. Comput Methods Programs Biomed. 2013;110(1):76–81.

76. Stüttgen G, Flesch U, Witt H, Wendt H. Thermographic analysis of skin test reaction using AGA thermovision. Arch Dermatol Res. 1980;268(2):113–28.

77. Flesch U, Wegener OH, Scheffler A, Ernst H. Thermometry of the surface of human skin. A study on a model using thermocouples, thermistors, thermovision and thermodyes. Phys Med Biol. 1976;21(3):422–8.

78. Oster A. Thermography and thermovision. Infrared irradiation of the skin used for the graphic registration of skin temperature. Ugeskr Laeger. 1966;128(35):1016–8.

79. Gao Y, Tian GY, Wang P, Wang H, Gao B, Woo WL, Li K. Electromagnetic pulsed thermography for natural cracks inspection. Sci Rep. 2017;7(7):42073. doi:10.1038/srep42073.

80. Voiticovschi-Iosob C, Antonaci F, Rossi E, Costa A, Pucci E, Sances G, Dalla Volta G. Frontal thermography in healthy individuals and headache patients: reliability of the method. J Headache Pain. 2015;16(Suppl 1):A122. doi:10.1186/1129-2377-16-S1-A122.

81. Liu B, Zhang H, Fernandes H, Maldague X et al. Correction: quantitative evaluation of pulsed thermography, lock-in thermography and vibrothermography on foreign object defect (FOD) in CFRP. Sensors 2016, 16, doi:10.3390/s16050743. Sensors (Basel). 2017 Jan 20;17(1). pii: E195.

82. Hoffmann N, Weidner F, Urban P, Meyer T, Schnabel C, Radev Y, Schackert G, Petersohn U, Koch E, Gumhold S, Steiner G, Kirsch M. Framework for 2D-3D image fusion of infrared thermography with preoperative MRI. Biomed Tech (Berl). 2017 Jan 23. pii: /j/bmte.ahead-of-print/bmt-2016-0075/bmt-2016-0075.xml.

83. Sagaidachnyi AA, Fomin AV, Usanov DA, Skripal AV. Thermography-based blood flow imaging in human skin of the hands and feet: a spectral filtering approach. Physiol Measur. 2017;38(2):272–88.

84. Sathiyabarathi M, Jeyakumar S, Manimaran A, Pushpadass HA, Sivaram M, Ramesha KP, Das DN, Kataktalware MA, Jayaprakash G, Patbandha TK. Investigation of body and udder skin surface temperature differentials as an early indicator of mastitis in Holstein Friesian crossbred cows using digital infrared thermography technique. Vet World. 2016;9(12):1386–91.

85. Sun G, Nakayama Y, Dagdanpurev S, Abe S, Nishimura H, Kirimoto T, Matsui T. Remote sensing of multiple vital signs using a CMOS camera-equipped infrared thermography system and its clinical application in rapidly screening patients with suspected infectious diseases. Int J Infect Dis. 2017;16(55):113–7.

86. Zanghi BM. Eye and ear temperature using infrared thermography are related to rectal temperature in dogs at rest or with exercise. Front Vet Sci. 2016;19(3):111.

87. Alexandre D, Prieto M, Beaumont F, Taiar R, Polidori G. Wearing lead aprons in surgical operating rooms: ergonomic injuries evidenced by infrared thermography. J Surg Res. 2016;2(209):227–33.

88. Thompson CL, Scheidel C, Glander KE, Williams SH, Vinyard CJ. An assessment of skin temperature gradients in a tropical primate using infrared thermography and subcutaneous implants. J Therm Biol. 2017;63:49–57.

89. Holm JK, Kellett JG, Jensen NH, Hansen SN, Jensen K, Brabrand M. Prognostic value of infrared thermography in an emergency department. Eur J Emerg Med. 2016 Dec 20.

90. Deery DM, Rebetzke GJ, Jimenez-Berni JA, James RA, Condon AG, Bovill WD, Hutchinson P, Scarrow J, Davy R, Furbank RT. Methodology for high-throughput field phenotyping of canopy temperature using airborne thermography. Front Plant Sci. 2016;6(7):1808.

91. Liu X, Huang L, Guo D, Xie G. Infrared thermography investigation of an evaporating water/oil meniscus in confined geometry. Langmuir. 2017;33(1):197–205. doi:10.1021/acs.langmuir.6b03482.

92. Xie J, Hsieh SJ, Wang HJ, Tan Z. Thermography and machine learning techniques for tomato freshness prediction. Appl Opt. 2016;55(34):D131–9.

93. Ludwig N, Trecroci A, Gargano M, Formenti D, Bosio A, Rampinini E, Alberti G. Thermography for skin temperature evaluation during dynamic exercise: a case study on an incremental maximal test in elite male cyclists. Appl Opt. 2016;55(34):D126–30.

94. Maierhofer C, Röllig M, Krankenhagen R, Myrach P. Comparison of quantitative defect characterization using pulse-phase and lock-in thermography. Appl Opt. 2016;55(34): D76–86.

95. Skala J, Svantner M, Tesar J, Franc A. Active thermography inspection of protective glass contamination on laser scanning heads. Appl Opt. 2016;55(34):D60–6.

96. Tokumitsu J, Kawata S, Ichioka Y, Suzuki T. Adaptive binarization using a hybrid image processing system. Appl Opt. 1978;17(16):2655–7.

97. Auer M, Regitnig P, Stollberger R, Ebner F, Holzapfel GA. A methodology to study the morphologic changes in lesions during in vitro angioplasty using MRI and image processing. Med Image Anal. 2008;12(2):163–73.

98. Kondo T, Ikeda T, Go J. Image processing and quantitative assessment in computer tomography for non-surgical treatment of brain tumours. Acta Neurochir (Wien). 1982;64 (1–2):19–37.

99. Mäder U, Quiskamp N, Wildenhain S, Schmidts T, Mayser P, Runkel F, Fiebich M. Image-processing scheme to detect superficial fungal infections of the skin. Comput Math Methods Med. 2015;2015:851014.

100. Singh A, Dutta MK, ParthaSarathi M, Uher V, Burget R. Image processing based automatic diagnosis of glaucoma using wavelet features of segmented optic disc from fundus image. Comput Methods Programs Biomed. 2016;124:108–20.

101. Scolozzi P, Jaques B. Intraoral open reduction and internal fixation of displaced mandibular angle fractures using a specific ad hoc reduction-compression forceps: a preliminary study. Oral Surg Oral Med Oral Pathol Oral Radiol Endod. 2008;106(4):497–501.

102. Report of the Ad Hoc Committee on Physician-Specific Mortality Rates for Cardiac Surgery. Ann Thorac Surg. 1993 Nov;56(5):1200–2.

103. Ruiz N. Filling holes in peptidoglycan biogenesis of *Escherichia coli*. Curr Opin Microbiol. 2016;34:1–6.

104. Schmidt L, Danziger S. Filling holes in the safety net? Material hardship and subjective well-being among disability benefit applicants and recipients after the 1996 welfare reform. Soc Sci Res. 2012;41(6):1581–97.

105. Wu S, Tao A, Jiang H, Xu Z, Perez V, Wang J. Vertical and horizontal corneal epithelial thickness profile using ultra-high resolution and long scan depth optical coherence tomography. PLoS One. 2014;9(5):e97962.

106. Jakobsone G, Stenvik A, Espeland L. Importance of the vertical incisor relationship in the prediction of the soft tissue profile after Class III bimaxillary surgery. Angle Orthod. 2012;82(3):441–7.

107. Michiels G, Sather AH. Validity and reliability of facial profile evaluation in vertical and horizontal dimensions from lateral Cephalograms and lateral photographs. Int J Adult Orthodon Orthognath Surg. 1994;9(1):43–54.

108. Koprowski R. Image analysis for ophthalmological diagnosis. Image processing of Corvis® ST images using Matlab®. Springer 2016.

109. Koprowski R. Processing of hyperspectral medical images, applications in dermatology using Matlab®. Springer 2017.

110. Delorme T, Garcia A, Nasr P. A longitudinal analysis of methicillin-resistant and sensitive *Staphylococcus aureus* incidence in respect to specimen source, patient location, and temperature variation. Int J Infect Dis. 2017;54:50–7. doi:10.1016/j.ijid.2016.11.405.

111. Markota A, Košir AS, Balažič P, Živko I, Sinkovič A. A novel esophageal heat transfer device for temperature management in an adult patient with severe meningitis. J Emerg Med. 2017;52(1):e27–8. doi:10.1016/j.jemermed.2016.07.086.

112. Peters C, Rosch RE, Hughes E, Ruben PC. Temperature-dependent changes in neuronal dynamics in a patient with an SCN1A mutation and hyperthermia induced seizures. Sci Rep. 2016;1(6):31879.

113. Chiang YZ, Al-Niaimi F, Madan V. Comparative efficacy and patient preference of topical anaesthetics in dermatological laser treatments and skin microneedling. J Cutan Aesthet Surg. 2015 Jul–Sep;8(3):143–6.

114. Vazquez-Martinez O, Eichelmann K, Garcia-Melendez M, Miranda I, Avila-Lozano A, Vega D, Ocampo-Candiani J. Pulsed dye laser for early treatment of scars after dermatological surgery. J Drugs Dermatol. 2015;14(11):1209–12.

115. Albuquerque RS, Mariani C Neto, Bersusa AA, Dias VM, Silva MI. Newborns' temperature submitted to radiant heat and to the Top Maternal device at birth. Rev Lat Am Enfermagem. 2016 Aug 8;24:e2741. doi: 10.1590/1518-8345.0305.2741. English, Portuguese, Spanish.

116. Bueno C, Menna-Barreto L. Development of sleep/wake, activity and temperature rhythms in newborns maintained in a neonatal intensive care unit and the impact of feeding schedules. Infant Behav Dev. 2016;44:21–8.

117. Sharma D, Pandita A, Murki S, Pratap OT. Administration of inhaled gases at a temperature of 33.5 °C versus 37 °C for ventilated asphyxiated newborns undergoing therapeutic hypothermia. Paediatr Child Health. 2015 Aug–Sep;20(6):296.

118. Boisnic S, Divaris M, Branchet MC, Nelson AA. Split-face histological and biochemical evaluation of tightening efficacy using temperature- and impedance-controlled continuous non-invasive radiofrequency energy. J Cosmet Laser Ther. 2017;6:1–5.

119. Wu HC, Kumar A, Wang J, Bi XF, Tomé CN, Zhang Z, Mao SX. Rolling-induced face centered cubic titanium in hexagonal close packed titanium at room temperature. Sci Rep. 2016;12(6):24370.

120. Cai C, Yin X, He S, Jiang W, Si C, Struik PC, Luo W, Li G, Xie Y, Xiong Y, Pan G. Responses of wheat and rice to factorial combinations of ambient and elevated CO_2 and temperature in FACE experiments. Glob Chang Biol. 2016;22(2):856–74.

121. Morrison SA, Ciuha U, Zavec-Pavlinić D, Eiken O, Mekjavic IB. The effect of a live-high train-high exercise regimen on behavioural temperature regulation. Eur J Appl Physiol. 2017;117(2):255–65.

122. Wu Y, Nieuwenhoff MD, Huygen FJ, van der Helm FC, Niehof S, Schouten AC. Characterizing human skin blood flow regulation in response to different local skin temperature perturbations. Microvasc Res. 2016;21(111):96–102.

123. He J, Zhu X. Combining improved gray-level co-occurrence matrix with high density grid for myoelectric control robustness to electrode shift. IEEE Trans Neural Syst Rehabil Eng. 2016 Dec 22.

124. Zhang J, Wang G, Feng Y, Sa Y. Comparison of contourlet transform and gray level co-occurrence matrix for analyzing cell-scattered patterns. J Biomed Opt. 2016;21 (8):86013.

125. Rajković N, Kolarević D, Kanjer K, Milošević NT, Nikolić-Vukosavljević D, Radulovic M. Comparison of Monofractal, Multifractal and gray level Co-occurrence matrix algorithms in analysis of Breast tumor microscopic images for prognosis of distant metastasis risk. Biomed Microdevices. 2016;18(5):83.

126. Pantic I, Dimitrijevic D, Nesic D, Petrovic D. Gray level co-occurrence matrix algorithm as pattern recognition biosensor for oxidopamine-induced changes in lymphocyte chromatin architecture. J Theor Biol. 2016;7(406):124–8.

127. Bunnell WP. An objective criterion for scoliosis screening. J Bone Joint Surg. 1984;66A:1381–7.

128. Dua S, Kandiraju N, Chowriappa P. Region quad-tree decomposition based edge detection for medical images. Open Med Inform J. 2010;28(4):50–7.

129. Baldi A. Two-dimensional phase unwrapping by quad-tree decomposition. Appl Opt. 2001;40(8):1187–94.

130. Han L, Long T, Tang W, Liu L, Jing W, Tian WD, Long J. Correlation between condylar fracture pattern after parasymphyseal impact and condyle morphological features: a retrospective analysis of 107 Chinese patients. Chin Med J (Engl). 2017 Feb 20;130 (4):420–27.

131. Carnacina I, Larrarte F, Leonardi N. Acoustic measurement and morphological features of organic sediment deposits in combined sewer networks. Water Res. 2017;1(112):279–90. doi:10.1016/j.watres.2017.01.050.

132. Ko PS, Liu YC, Yeh CM, Gau JP, Yu YB, Hsiao LT, Tzeng CH, Chen PM, Chiou TJ, Liu CJ, Liu JH. The uniqueness of morphological features of pure erythroid leukemia in myeloid neoplasm with erythroid predominance: a reassessment using criteria revised in the 2016 World Health Organization classification. PLoS One. 2017;12(2):e0172029.

133. Koprowski R. The use of image processing methods of thermography in the diagnosis of lateral spinal curvatures. Silesian University of Technology, doctoral thesis, 2003.

134. Hsiao CY, Sung HC, Hu S, Huang CH. Skin pretreatment with conventional non-fractional ablative lasers promote the transdermal delivery of tranexamic acid. Dermatol Surg. 2016;42(7):867–74.

135. Jagdeo JR, Brody NI, Spandau DF, Travers JB. Important implications and new uses of ablative lasers in dermatology: fractional carbon dioxide laser prevention of skin cancer. Dermatol Surg. 2015;41(3):387–9.

136. Lipozenčić J, Mokos ZB. Will nonablative rejuvenation replace ablative lasers? Facts and controversies. Clin Dermatol. 2013 Nov–Dec;31(6):718–24.

137. Ciocon DH1, Doshi D, Goldberg DJ. Non-ablative lasers. Curr Probl Dermatol. 2011;42:48–55.

138. Karaman M, Gün T, Temelkuran B, Aynacı E, Kaya C, Tekin AM. Comparison of fiber delivered CO_2 laser and electrocautery in transoral robot assisted tongue base surgery. Eur Arch Otorhinolaryngol. 2017 Feb 11.

139. Sokol ER, Karram MM. Use of a novel fractional CO_2 laser for the treatment of genitourinary syndrome of menopause: 1-year outcomes. Menopause. 2017 Feb 6.

140. Paulos RS, Seino PY, Fukushima KA, Marques MM, de Almeida FC, Ramalho KM, de Freitas PM, Brugnera A Junior, Moreira MS. Effect of Nd:YAG and CO_2 laser irradiation on prevention of enamel demineralization in orthodontics: in vitro study. Photomed Laser Surg. 2017 Jan 18.

141. Bertolin A, Lionello M, Salis G, Rizzotto G, Lucioni M. Two-stage CO_2-laser-assisted bilateral cordectomy for cT1b glottic carcinoma. Am J Otolaryngol. 2017 Jan 18. pii: S0196-0709(16)30292–7.

142. Jiang X, Zhou Z, Ding X, Deng X, Zou L, Li B. Level set based Hippocampus segmentation in MR images with improved initialization using region growing. Comput Math Methods Med. 2017;2017:5256346. doi:10.1155/2017/5256346.

143. Tian Y, Pan Y, Duan F, Zhao S, Wang Q, Wang W. Automated segmentation of coronary arteries based on statistical region growing and heuristic decision method. Biomed Res Int. 2016;2016:3530251.

144. Li W, Fontanelli O, Miramontes P. Size distribution of function-based human gene sets and the split-merge model. R Soc Open Sci. 2016;3(8):160275.

145. Calus MP, Bouwman AC, Schrooten C, Veerkamp RF. Efficient genomic prediction based on whole-genome sequence data using split-and-merge Bayesian variable selection. Genet Sel Evol. 2016;48(1):49.

146. Huynh HT, Le-Trong N, Bao PT, Oto A, Suzuki K. Fully automated MR liver volumetry using watershed segmentation coupled with active contouring. Int J Comput Assist Radiol Surg. 2017;12(2):235–43. doi:10.1007/s11548-016-1498-9.

147. Rashwan S, Sarhan A, Faheem MT, Youssef BA. Fuzzy watershed segmentation algorithm: an enhanced algorithm for 2D gel electrophoresis image segmentation. Int J Data Min Bioinform. 2015;12(3):275–93.

148. Chen J, Alber MS, Chen DZ. A hybrid approach for segmentation and tracking of *Myxococcus xanthus* swarms. IEEE Trans Med Imaging. 2016 Mar 30.

149. Badoual A, Schmitter D, Uhlmann V, Unser M. Multiresolution subdivision snakes. IEEE Trans Image Process. 2017;26(3):1188–201.

150. Bonanno L, Sottile F, Ciurleo R, Di Lorenzo G, Bruschetta D, Bramanti A, Ascenti G, Bramanti P, Marino S. Automatic algorithm for segmentation of atherosclerotic carotid plaque. J Stroke Cerebrovasc Dis. 2017;26(2):411–6.

151. Zhang R, Zhu S, Zhou Q. A novel gradient vector flow snake model based on convex function for infrared image segmentation. Sensors (Basel). 2016 Oct 21;16(10). pii: E1756.

152. Piorkowski A, Nurzynska K, Gronkowska-Serafin J, Selig B, Boldak C, Reska D. Influence of applied corneal endothelium image segmentation techniques on the clinical parameters. Comput Med Imaging Graph. 2017;55:13–27.

153. Kim YJ, Lee SH, Park CM, Kim KG. Evaluation of semi-automatic segmentation methods for persistent ground glass nodules on thin-section CT scans. Healthc Inform Res. 2016;22 (4):305–15.

154. Renauld E, Descoteaux M, Bernier M, Garyfallidis E, Whittingstall K. Semi-automatic segmentation of optic radiations and LGN, and their relationship to EEG alpha waves. PLoS One. 2016;11(7):e0156436.

155. Hamel P, Falinski K, Sharp R, Auerbach DA, Sánchez-Canales M, Dennedy-Frank PJ. Sediment delivery modeling in practice: Comparing the effects of watershed characteristics and data resolution across hydroclimatic regions. Sci Total Environ. 2016 Dec 28. pii: S0048-9697(16)32791–7.

156. McCarney-Castle K, Childress TM, Heaton CR. Sediment source identification and load prediction in a mixed-use Piedmont watershed, South Carolina. J Environ Manag. 2016 Oct 28. pii: S0301-4797(16)30820–9.

157. Potter TL, Coffin AW. Assessing pesticide wet deposition risk within a small agricultural watershed in the Southeastern Coastal Plain (USA). Sci Total Environ. 2016 Dec 24. pii: S0048-9697(16)32451–2.

158. Christenson EC, Serre ML. Integrating remote sensing with nutrient management plans to calculate nitrogen parameters for swine CAFOs at the sprayfield and sub-watershed scales. Sci Total Environ. 2016 Dec 22. pii: S0048-9697(16)32718–8.

159. Smith MB, Chaigne A, Paluch EK. An active contour ImageJ plugin to monitor daughter cell size in 3D during cytokinesis. Methods Cell Biol. 2017;137:323–40.

160. Chao M, Wei J, Yuan Y, Lo Y. SU-G-BRA-10: marker free lung tumor motion tracking by an active contour model on cone beam CT projections for stereotactic body radiation therapy of lung cancer. Med Phys. 2016;43(6):3637.

161. Joseph S, Adnan A, Adlam D. Automatic segmentation of coronary morphology using transmittance-based lumen intensity-enhanced intravascular optical coherence tomography images and applying a localized level-set-based active contour method. J Med Imaging (Bellingham). 2016;3(4):044001.

162. Hoogi A, Subramaniam A, Veerapaneni R, Rubin D. Adaptive estimation of active contour parameters using convolutional neural networks and texture analysis. IEEE Trans Med Imaging. 2016 Nov 11.

163. Gregoretti F, Cesarini E, Lanzuolo C, Oliva G, Antonelli L. An automatic segmentation method combining an active contour model and a classification technique for detecting polycomb-group proteinsin high-throughput microscopy images. Methods Mol Biol. 2016;1480:181–97. doi:10.1007/978-1-4939-6380-5_16.

164. Rebouças Filho PP, Cortez PC, da Silva Barros AC, Albuquerque VHC, Tavares JMR. Novel and powerful 3D adaptive crisp active contour method applied in the segmentation of CT lung images. Med Image Anal. 2017;35:503–16.

165. Wang J, Zhao S, Liu Z, Tian Y, Duan F, Pan Y. An active contour model based on adaptive threshold for extraction of cerebral vascular structures. Comput Math Methods Med. 2016;2016:6472397.

166. Zhuang M, Dierckx RA, Zaidi H. Generic and robust method for automatic segmentation of PET images using an active contour model. Med Phys. 2016;43(8):4483.

167. Ghayoumi Zadeh H, Haddadnia J, Montazeri A. A model for diagnosing breast cancerous tissue from thermal images using active contour and Lyapunov exponent. Iran J Publ Health. 2016;45(5):657–69.

168. Mostaar A, Houshyari M, Badieyan S. A novel active contour model for MRI brain segmentation used in radiotherapy treatment planning. Electron Physician. 2016;8(5): 2443–51.

169. Zareei A, Karimi A. Liver segmentation with new supervised method to create initial curve for active contour. Comput Biol Med. 2016;1(75):139–50.

170. Cohen LD, Cohen I. Finite-element method for active contour models and balloons for 2D and 3D images. IEEE Trans Pattern Anal Mach Intell. 1993;15(11):1131–47.

171. Yang Q, Zhang J, Nozato K, Saito K, Williams DR, Roorda A, Rossi EA. Closed-loop optical stabilization and digital image registration in adaptive optics scanning light ophthalmoscopy. Biomed Opt Express. 2014;5(9):3174–91.

172. Su JY, Ting SC, Chang YH, Yang JT. Aerodynamic trick for visual stabilization during downstroke in a hovering bird. Phys Rev E Stat Nonlinear Soft Matter Phys. 2011;84 (1 Pt 1):012901.

173. Zhai B, Zheng J, Li B. Digital image stabilization based on adaptive motion filtering with feedback correction. Multimedia Tools Appl. 2016;75(19):12173–2200.

174. Shukla D, Jha RK, Ojha A. Digital image stabilization using similarity transformation over constrained differential-radon warping vectors. Sig Process Image Commun. 2016;47: 115–30.

175. Unser M, Sage D, Van De Ville D. Multiresolution monogenic signal analysis using the Riesz-Laplace wavelet transform. IEEE Trans Image Process. 2009;18(11):2402–18.

176. Xue B, Liu Y, Cui L, Bai X, Cao X, Zhou F. Video stabilization in atmosphere turbulent conditions based on the Laplacian-Riesz pyramid. Opt Express. 2016;24(24):28092–103. doi:10.1364/OE.24.028092.

177. Cirujeda P, Dicente Cid Y, Muller H, Rubin D, Aguilera TA, Loo BW, Diehn M, Binefa X, Depeursinge A. A 3-D Riesz-covariance texture model for prediction of nodule recurrence in lung CT. IEEE Trans Med Imaging. 2016 July 18.

178. Wu GC, Baleanu D, Xie HP. Riesz Riemann-Liouville difference on discrete domains. Chaos. 2016;26(8):084308.

179. Cirujeda P, Cid Y, Muller H, Rubin D, Aguilera T, Loo B, Diehn M, Binefa X, Depeursinge A. A 3-D Riesz-covariance texture model for prediction of nodule recurrence in lung CT. IEEE Trans Med Imaging. 2016 July 18.

180. Cirujeda P, Muller H, Rubin D, Aguilera TA, Loo BW, Diehn M, Binefa X, Depeursinge A. 3D Riesz-wavelet based covariance descriptors for texture classification of lung nodule tissue in CT. Conf Proc IEEE Eng Med Biol Soc. 2015;2015:7909–12.

181. Koprowski R. Processing of hyperspectral medical images, applications in dermatology using Matlab®. Springer 2017.

182. Koprowski Robert, Wilczyński Sławomir, Wróbel Zygmunt, Błońska-Fajfrowska Barbara. Calibration and segmentation of skin areas in hyperspectral imaging for the needs of dermatology. BioMed Eng Online. 2014;13:113.

183. Koprowski R, Wilczyński S, Wróbel Z, Kasperczyk S, Błońska-Fajfrowska B. Automatic method for the dermatological diagnosis of selected hand skin features in hyperspectral imaging. Biomed Eng Online. 2014;13:47.

184. Koprowski R. Hyperspectral imaging in medicine: image pre-processing problems and solutions in Matlab. J Biophotonics. 2015;8(11–12):935–43. doi:10.1002/jbio.201400133 Epub 2015 Feb 9.

185. Sonka M, Michael Fitzpatrick J. Medical image processing and analysis. In: Handbook of Medical, Imaging. Bellingham: SPIE; 2000.

186. Koprowski R. Quantitative assessment of the impact of biomedical image acquisition on the results obtained from image analysis and processing. BioMed Eng Online. 2014;13:93.

187. Gonzalez R, Woods R. Digital image processing. New York: Addison-Wesley Publishing Company; 1992.

188. Koprowski R, Wróbel Z. Image processing in optical coherence tomography using Matlab. Katowice: Wydawnictwo Uniwersytet Śląski; 2011.

189. Noguchi N, Okuchi T. A Peltier cooling diamond anvil cell for low-temperature Raman spectroscopic measurements. Rev Sci Instrum. 2016;87(12):125107. doi:10.1063/1.4972254.

190. Daimon S, Iguchi R, Hioki T, Saitoh E, Uchida KI. Thermal imaging of spin Peltier effect. Nat Commun. 2016;12(7):13754. doi:10.1038/ncomms13754.

191. Buse B, Kearns S, Clapham C, Hawley D. Decontamination in the electron probe microanalysis with a Peltier-cooled cold finger. Microsc Microanal. 2016;22(5):981–6.

192. Otsu N. A threshold selection method from gray-level histograms. IEEE Trans Syst Man Cyber. 1979;9(1):62–6.

193. Abiramalatha T, Santhanam S, Mammen JJ, Rebekah G, Shabeer MP, Choudhury J, Nair SC. Utility of neutrophil volume conductivity scatter (VCS) parameter changes as sepsis screen in neonates. J Perinatol. 2016;36(9):733–8. doi:10.1038/jp.2016.69.

194. Kim SJ, Choi CG, Kim JK, Yun SC, Jahng GH, Jeong HK, Kim EJ. Effects of MR parameter changes on the quantification of diffusion anisotropy and apparent diffusion coefficient in diffusion tensor imaging: evaluation using a diffusional anisotropic phantom. Korean J Radiol. 2015;16(2):297–03.

195. Draganski B, Ashburner J, Hutton C, Kherif F, Frackowiak RS, Helms G, Weiskopf N. Regional specificity of MRI contrast parameter changes in normal ageing revealed by voxel-based quantification (VBQ). Neuroimage. 2011;55(4):1423–34.

196. Nijsten SM, van Elmpt WJ, Mijnheer BJ, Minken AW, Persoon LC, Lambin P, Dekker AL. Prediction of DVH parameter changes due to setup errors for breast cancer treatment based on 2D portal dosimetry. Med Phys. 2009;36(1):83–94.

197. Magosso E, Ursino M. Effects of cardiovascular parameter changes on heart rate variability: analysis by a mathematical model of short-term cardiovascular regulation. Conf Proc IEEE Eng Med Biol Soc. 2004;6:3905–8.

198. Koprowski R, Korzyńska A, Wróbel Z, Zieleźnik W, Witkowska A, Małyszek J, Wójcik W. Influence of the measurement method of features in ultrasound images of the thyroid in the diagnosis of Hashimoto's disease. Biomed Eng Online. 2012;28(11):91.

199. Koprowski R. Quantitative assessment of the impact of biomedical image acquisition on the results obtained from image analysis and processing. Biomed Eng Online. 2014;4(13):93.

200. Byrski A, Kisiel-Dorohinicki M. Agent-based model and computing environment facilitating the development of distributed computational intelligence systems. Lecture Notes in Computer Science (including subseries Lecture Notes in Artificial Intelligence and Lecture Notes in Bioinformatics), 5545 LNCS (PART 2), 2009. p. 865–74.

201. Krzywicki D, Faber L, Byrski A, Kisiel-Dorohinicki M. Computing agents for decision support systems. Future Gener Comput Syst. 2014;37:390–400.

202. Byrski A, Schaefer R, Smolka M, Cotta C. Asymptotic guarantee of success for multi-agent memetic systems. Bull Pol Acad Sci Tech Sci. 2013;61(1):257–78.

203. Suphachokauychai S, Sananpanich K. Detection of perforators using smartphone thermal imaging. Plast Reconstr Surg Glob Open. 2016 May 26;4(5):e722.

204. Foster KR, Morrissey JJ. Thermal aspects of exposure to radiofrequency energy: report of a workshop. Int J Hyperth. 2011;27(4):307–19.

205. Foster KR, Zhang H, Osepchuk JM. Document thermal response of tissues to millimeter waves: implications for setting exposure guidelines. Health Phys. 2010;99(6):806–10.

206. Foster KR. Thermographic detection of breast cancer source of the document. IEEE Eng Med Biol Mag. 1998;17(6):10.

The manufacturer's authorised representative in the EU is Springer
Nature Customer Service Centre GmbH, Europaplatz 3, 69115 Heidelberg,
Germany. If you have any concerns regarding our products, please
contact ProductSafety@springernature.com

Printed and bound by CPI Group (UK) Ltd, Croydon, CR0 4YY
01/12/2025
02008619-0011